Summer Counts!

Grade 6

Acknowledgments
Product Development: Margaret Fetty
Editor: Michelle Howell
Cover Illustration: Tad Herr
Design and Production: Creative Pages, Inc.
Production Supervision: Sandy Batista

ISBN 1-59137-118-X
Options Publishing Inc.
P.O. Box 1749
Merrimack, NH 03054-1749
www.optionspublishing.com
Phone: 800-782-7300 Fax: 866-424-4056

15 14 13 12 11 10 9 8 7 6 5 4

Dear Parent,

Summer is a time for relaxing and having fun. It can also be a time for learning. *Summer Counts!* can help improve your child's understanding of important skills learned in the past school year while preparing him or her for the year ahead.

Summer Counts! provides grade-appropriate practice in subjects such as reading, language arts, vocabulary, and math. The ten theme-related chapters include activities and puzzles to motivate your child throughout the summer.

When working through the book, encourage your child to share his or her learning with you. You may want to tear out the answer key at the back of the book and use it to check your child's progress. With *Summer Counts!* your child will discover that learning happens anytime—even in the summer!

Apreciados padre,

El verano es una época para descansar y divertirse. También puede ser una época para aprender. *Summer Counts!* puede ayudar a que su hijo(a) mejore las destrezas importantes que aprendió el pasado año escolar al mismo tiempo que lo(a) prepara para el año que se aproxima.

Summer Counts! provee la práctica apropiada para cada grado en las asignaturas como la lectura, las artes del lenguaje y las matemáticas. Los diez capítulos temáticos incluyen actividades y rompecabezas que motivarán a su hijo(a) durante el verano.

Cuando trabaje con el libro, anime a su hijo(a) a que comparta lo que ha aprendido con Ud. Si Ud. desea puede desprender la página de las respuestas que aparece en la parte trasera del libro. Puede usar la misma para revisar el progreso de su hijo(a). ¡Con *Summer Counts!* su hijo(a) descubrirá que el aprendizaje puede ocurrir en cualquier momento—inclusive en el verano!

Table of Contents

CHAPTER 1

First Night on the Job

On Monday night, I was eating dinner with my family when my aunt called. My aunt owns a restaurant in town where my family and I go for special occasions. My mother answered the phone and spoke with her for five minutes before saying, "I'll go tell her right now. See you Friday."

My mother sat back down at the table and turned to me. "Keisha," she said, "your aunt says they're short-staffed at the restaurant this weekend. She needs more people and asked if you could work."

How on earth could I work in a restaurant if I didn't know how to cook? I didn't want to disappoint my aunt, or my mother, so I agreed. Later that evening I could have kicked myself for ever agreeing to work there. As the night went on, I thought more and more about the job. I had never even seen the kitchen in my aunt's restaurant, and I was expected to work there!

That week, I woke up every morning with butterflies in my stomach, thinking about Friday night. I thought about telling my aunt that I was sick and would not be able to work after all—but I didn't want to lie.

Thursday night after dinner, I approached my mother. "Mom," I said, "I'm not sure if I really want to work at the restaurant tomorrow."

My mother turned to me and smiled. "You're just nervous because this is your first job. It's natural to be nervous, but your aunt will explain everything to you tomorrow. You will be fine, trust me."

Friday arrived and the butterflies in my stomach were worse than ever. I was so worried about the job that it was hard to pay attention at school.

The bus dropped me off at the restaurant at 4:30—right before the dinner shift started. My aunt greeted me outside the restaurant. As we walked around to the back door, she explained my responsibility for the night. It would be my job to pick up all the dirty dishes and deliver them to Eddie, the dishwasher. My aunt showed me the gray, plastic bins where I would put the dirty dishes and introduced me to Eddie.

It took a few tries to learn to stack the dishes in the bins properly. Once I figured it out, I was running back and forth handing Eddie bin after bin of dirty dishes. As I was putting my apron away, my aunt appeared.

"So Keisha," she said with a broad smile on her face, "how was your first night on the job?"

"Well," I said, "it wasn't as bad as I expected." When I stopped and thought for a moment, I felt silly for worrying. "I really liked my first job!"

"Great!" my aunt said, smiling. "You did a good job, and I need you to work again tomorrow night!"

First Night on the Job

Directions Use what you have read to answer the questions.

1. What is this selection mostly about?

   ____

   ____

2. Why is Keisha worried about going to work?

   ____

   ____

3. What happens after Keisha's mother gets off the phone?

   ____

   ____

4. Why did Keisha's aunt ask her to work the next night?

   ____

   ____

5. What is the author's purpose in writing this story?

   ____

   ____

6. What do you think Keisha's job will be the next night she works?

   ____

   ____

Pronouns: Object, Subject, and Possessive

REMEMBER

A **noun** is a word that names a person, place, or thing. A **pronoun** takes the place of a noun. **Subject pronouns** are *you, he, she, it, we,* and *they*. **Object pronouns** are *me, you, him, her, it, us,* and *them*. **Possessive** pronouns are *my, your, his, her, its, yours, their,* and *theirs*.

EXAMPLES

Subject pronoun: *He* asked Katie to watch his dogs.
Object pronoun: Katie walked *them* twice a day.
Possessive pronoun: Katie found *their* bowls under the table.

Making Money

Directions **Complete each sentence using the correct pronoun.**

1. Katie wants to visit ____________________ aunt in New Mexico. (she, her, hers)
2. ____________________ hopes to save enough money by June. (Her, They, She)
3. Alex is saving ____________________ money for a guitar. (his, he, him)
4. Katie and Alex think that ____________________ pet-sitting business will be successful. (they, their, theirs)
5. Mrs. Able asked ____________________ to watch her cats, Bert and Ernie. (them, they, their)
6. The cats spend most of ____________________ time indoors. (theirs, they, their)
7. "Alex and ____________________ will take good care of your cats," Katie told Mrs. Able. (me, I, us)
8. We will visit ____________________ twice a day," Katie said. (them, they, their)
9. "I know ____________________ cats will have the very best care," said Mrs. Able. (I, me, my)
10. Alex and Katie are happy to get paid to do a job ____________________ both enjoy. (their, they, theirs)

Money Matches

Directions Choose a word from the box below that matches each definition. Write it on the line.

Financial	guarantee	income	inexpensive	inherit
withdraw	savings	value	wage	

1. A promise or an assurance, usually in writing ____________________

2. A sum of money set aside for later use ____________________

3. Monetary or material worth ____________________

4. Reasonable in price ____________________

5. To be left money or property in a will ____________________

Financial Report

Directions Read the report. Use words from the box above to fill in the blanks.

The **(6)** ____________________ Report for Pet-Sitting Friends

This year we were able to double our annual **(7)** ________________ by increasing our hourly **(8)** ________________ per client. We deposited this additional money into our **(9)** ________________ account in the bank for later use. We did have to **(10)** ____________________ money from the account to pay for dog biscuits to reward our best clients.

Multiplying Fractions

REMEMBER

To multiply a fraction and a whole number:

- Multiply the numerator by the whole number.
- Keep the same denominator.
- Reduce the fraction to lowest terms.

EXAMPLES

$$\frac{1}{3} \times 2 = \frac{1 \times 2}{3} = \mathbf{\frac{2}{3}} \qquad \frac{1}{2} \times 6 = \frac{1 \times 6}{2} = \frac{6}{2} = \mathbf{3}$$

$$\frac{3}{4} \times 6 = \frac{3 \times 6}{4} = \frac{18}{4} = 4\frac{2}{4} = \mathbf{4\frac{1}{2}}$$

Fraction Times

Directions **Complete each multiplication problem. Reduce your answer to lowest terms. The first one is done for you.**

1. $\frac{1}{2} \times 1$ can be written $\frac{1 \times 1}{2} = \frac{1}{2}$

2. $\frac{2}{4} \times 1$ can be written $\frac{2 \times \_\_}{4} = \frac{\_\_}{4} = \_\_\_\_$

3. $\frac{2}{4} \times 2$ can be written $\frac{\_\_ \times \_\_}{4} = \frac{\_\_}{4} = \_\_\_\_$

4. $\frac{2}{3} \times 3$ can be written $\frac{\_\_ \times \_\_}{3} = \frac{\_\_}{3} = \_\_\_\_$

5. $\frac{3}{4} \times 8$ can be written $\frac{\_\_ \times \_\_}{4} = \frac{\_\_}{4} = \_\_\_\_$

6. $\frac{5}{12} \times 5$ can be written $\frac{\_\_ \times \_\_}{12} = \frac{\_\_}{12} = \_\_\_\_$

Making Money

Directions Zander was in charge of selling drinks at his school dance. He recorded the data below. Use the table to complete the questions below.

Drinks Sold at Dance				
Small Fruit Punch $0.50	Medium Fruit Punch $0.90	Large Fruit Punch $1.25	Small Juice $0.85	Medium Juice $1.50
卌 卌 卌 卌 卌 卌 卌 卌 |	卌 卌 卌 卌 卌 卌 卌 卌 卌 卌 卌 ||	卌 卌 ||||	卌 卌 卌 卌 卌 卌 卌 ||	卌 卌 |||

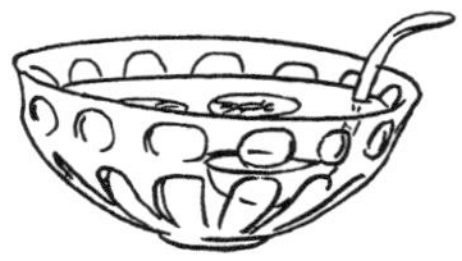

1. How many drinks were sold at each price?

a. $0.50 ________ **b.** $0.90 ________ **c.** $1.25 ________

d. $0.85 ________ **e.** $1.50 ________

2. How much total money was collected for each drink?

a. small fruit punch ________ **b.** medium fruit punch ________

c. large fruit punch ________ **d.** small juice ________

e. medium juice ________

3. Which drink had the median number of sales? ________________

4. What was the mean price of a drink? ________

5. Do you think that selling an extra-large fruit punch at $2.00 would have been a good idea? Explain your answer.

____

____

Mixed-Up Money

Directions **Read each definition. Unscramble the letters to solve. Write the words on the lines. Then use the numbered letters to solve the riddle.**

1. adtwwrhi To take money out of the bank

___ ___ ___ ___ ___ ___ ___ ___
(letter 4: 1)

2. boj A chore or a task

___ ___ ___
(letter 2: 2)

3. kanb A place to keep money

___ ___ ___ ___
(letter 3: 3)

4. roldal One hundred cents

___ ___ ___ ___ ___ ___
(letter 4: 4)

5. noyem Dollars and cents

___ ___ ___ ___ ___
(letter 4: 5; letter 5: 6)

6. esva To set aside or put away money

___ ___ ___ ___
(letter 2: 7)

7. densp To pay or use money

___ ___ ___ ___ ___
(letter 1: 8; letter 5: 9)

8. inocs Metal change

___ ___ ___ ___ ___
(letter 1: 10)

9. axt Money paid to the government

___ ___ ___
(letter 1: 11)

NOW SOLVE THE RIDDLE: Why did the skunk have to get a job?

___ ___   ___ ___ ___ ___   ___ ___ ___   ___ ___ ___   ___ ___ ___ ___ ___!
1 5   2 3 4 6   1 7 9   2 3 5   8 10 5 3 11

Job Posting

Suppose you wanted to earn money. What jobs could you do? Would you babysit, work in a store, or do yard work? Create a flyer advertising the job you want and the skills you have. Don't forget to include your business name and contact information.

CHAPTER 2

A Love for the World

The story of *Charlotte's Web*—one of the best-loved children's books ever written—begins on a farm inhabited by chickens, pigs, geese, sheep, spiders, rats, and people. The author, E. B. White, spent much of his life on a saltwater farm.

Elwin Brooks White was born in Mount Vernon, New York, in 1899. He won his first writing award at the age of nine for a poem about a pet mouse. After graduating from college, he went to work for *The New Yorker* magazine. White contributed essays and poems to the magazine for the rest of his life.

In the 1930s, E. B. White and his wife Katharine bought a small farm on the coast of Maine. White loved watching the cycle of life—birth, death, and rebirth—at close range on his farm. He was surrounded by animals large and small. He may have been feeling unhappy about a pig that had recently died in his care when he got an idea for a children's book. One morning, seated on a wooden bench in a boathouse on his farm, he began to write the story of a pig named Wilbur and the spider that saved his life.

White worked hard on the book, rewriting it eight times. He changed all the unnecessary words until he felt that it was written just right. He carefully researched the habits of spiders. While observing them, he saw that spiders were both skillful and useful. He decided they had gotten a "bad reputation" that they did not deserve.

Charlotte's Web was published in 1952. The book became an instant bestseller. The theme of the novel was friendship, courage, and the cycles of nature. White's story showed that joy can be found in everything, even tragedy.

A reader once asked White the meaning of his children's books. "All that I hope to say in books," he replied, "is that I love the world." E. B. White died in 1985, at the age of 86. Most of what he loved about the world can be found in *Charlotte's Web*.

A Love for the World

Directions **Using what you have just read, answer the questions.**

1. Who was E. B. White?

2. What was White's first job?

3. How did White get the idea for *Charlotte's Web*?

4. How did White feel about spiders?

5. Why do you think *Charlotte's Web* became an instant bestseller?

6. What does the final sentence of the article mean when it says: "Most of what he loved about the world can be found in *Charlotte's Web*"?

Nouns: Abstract and Concrete

REMEMBER

Abstract nouns name something you cannot see or touch. They express an idea, a quality, or a condition. Many abstract nouns are formed from other words by adding suffixes such as *-ness*, *-ship*, or *-ment*.

Concrete nouns name something you can see or touch. Most of the nouns you use every day are concrete nouns.

EXAMPLES

Concrete noun: *Beverly Cleary* is a well-known *author*.

Abstract noun: Cleary won *praise* for many of her books.

Concrete or Abstract?

Directions **Draw one line under each concrete noun and two lines under each abstract noun. Some sentences may have both.**

1. Beverly Cleary writes books for children and young adults.
2. She has won many awards.
3. Cleary writes about the problems and pleasures of growing up.
4. People ask Cleary how she gets ideas for her books.
5. Cleary says that she uses memories from her childhood.
6. She remembers how it felt to lose a tooth.
7. She also remembers her friendships.
8. She gets many letters from young readers.
9. Cleary is pleased that people enjoy her books.
10. She is grateful for her success.

Writing Tools

Directions **Choose a word from the box below that matches each definition. Write it on the line.**

conclusion	deadline	edit	essay	journal
opinion	persuade	read	rhymes	topic

1. A diary ____________________

2. A time limit ____________________

3. Lines of writing that have the same ending sounds ____________________

4. To check and correct a document ____________________

5. An article or piece of writing on a specific topic ____________________

Letter to the Editor

Directions **Read the letter. Use words from the box above to fill in the blanks.**

Dear Editor,

Enclosed is a book that I have written. The **(6)** ____________________ of my book is the environment. I hope it will **(7)** ____________________ readers to protect the environment. I am certain that my book will sway public **(8)** ____________________ about the dangers of cutting down the rain forests. I would like you to **(9)** ____________________ my book and consider it for print. In **(10)** ____________________, I'd like to thank you for taking the time to read it.

Sincerely,

Martin Harris

Probability

REMEMBER

Probability is about what might happen. Fractions can be used to tell how likely it is that something will happen.

EXAMPLE

Maria reaches into a bag that has 7 chips: 2 are red, 4 are green, and 1 is yellow. What is the probability that she will get a green chip?

$$P(g) = \frac{\text{number of green chips}}{\text{total number of chips}} = \frac{4}{7}$$

Read **P(g)** as **the probability of getting a green chip**.

The probability that Maria will get a green chip is $\frac{4}{7}$.

Directions **Use the information in the box above to answer the questions.**

1. What is the probability that Maria will get a red chip?

$$P(r) = \frac{\text{number of red chips}}{\text{total number of chips}} =$$

2. What is the probability that Maria will get a yellow chip?

$$P(y) = \frac{\text{number of yellow chips}}{\text{total number of chips}} =$$

3. What is the probability that Maria will get either a red chip or a yellow chip?

$$P(ry) = \frac{\text{number of red and yellow chips}}{\text{total number of chips}} =$$

4. What is the probability that Maria will get a blue chip?

$$P(b) = \frac{\text{number of blue chips}}{\text{total number of chips}} =$$

Don't Lose Your Marbles!

Directions **Yan has a bag containing 2 blue marbles and 2 red marbles. Use this information to complete the questions.**

1. Yan picks a marble out of the bag without looking. What is the probability that the marble is red? Write your answer as a fraction. __________

2. Next Yan picks 1 blue marble out of the bag. He puts it back in the bag and picks again. What's the probability that the second marble Yan picks will be blue? Write your answer as a fraction. __________

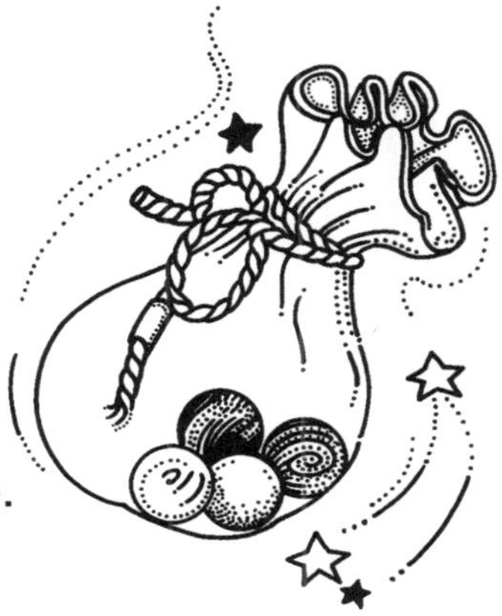

3. Yan puts the marble back in the bag so it has 2 blue and 2 red marbles again. Yan picks 1 red marble out of the bag and puts it aside. He does not replace the marble. Yan picks another marble out of the bag. What is the likelihood that the second marble Yan picks will also be red?

   a. It is more likely that the marble will be red than blue.

   b. It is equally likely that the marble will be either red or blue.

   c. It is more likely that the marble will be blue than red.

   Circle your answer above and explain.

   ____

   ____

Writer's Puzzle

Directions **Read each definition. Use words from the box to finish the puzzle.**

myth	letter	report	biography	limerick
poem	journal	haiku	mystery	folktale

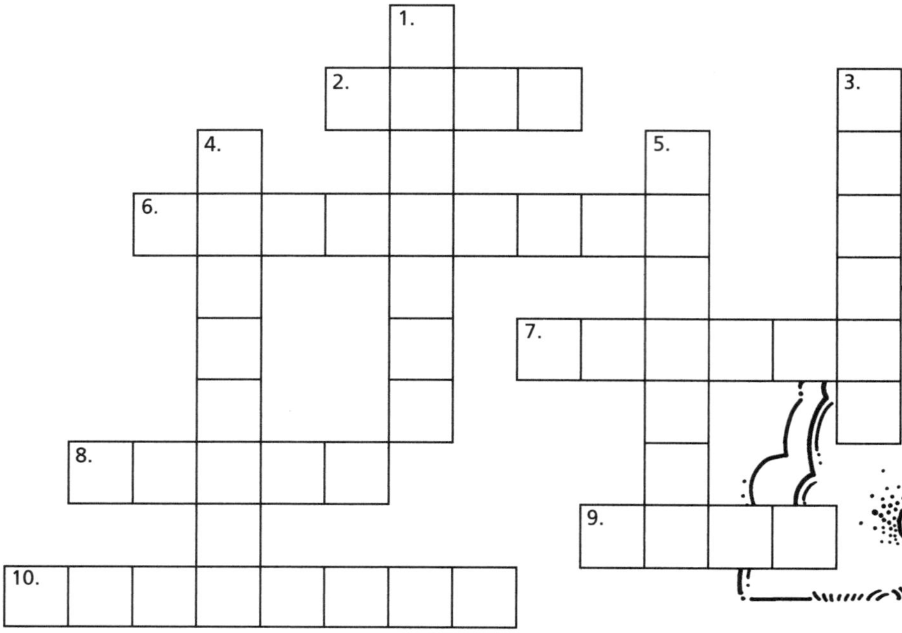

ACROSS

2. A short form of writing that may rhyme and can be used to express a feeling

6. A story about a person's life written by another person

7. A message written to someone

8. A Japanese poem that has three lines

9. A story that is used to explain how and why things happen in nature

10. A spoken story that explains how something came to be

DOWN

1. A diary

3. A detailed account or statement

4. A humorous poem with five lines

5. An unsolved crime story might be called this

Favorite Character

Do you have a favorite character from a story? What would it be like if this character came to visit you? Write a paragraph telling how you and the character would spend the day. Then draw a picture of you and the character doing one of these activities.

CHAPTER 3

Erh-Lang and the Seven Suns

Years ago in ancient China, seven suns ruled the skies. The suns were so proud that each wanted to be the brightest and hottest. Every day the suns rose one by one to show the people how big and beautiful they were.

As they worked the fields, the people sweated in the blinding glare from the seven suns. Slowly, the crops began to wither and the lakes and rivers began to dry up. Soon, the seven suns were up day and night competing to show who was the best. The people in China had no rest.

Finally, a wise woman said, "People, we must do something to prevent the suns from taking their toll on us. Let us consult the giant, Erh-Lang." Erh-Lang the giant was as strong as 50 men and could move mountains with his bare hands.

"Erh-Lang," cried the desperate people, "save us from these blazing suns." However, Erh-Lang just scratched his head and shrugged.

"I know," suggested the wise woman. "You could remove the top of a mountain peak and hide the suns inside."

Erh-Lang agreed, and as the next proud sun rose into the sky, Erh-Lang reached up and snatched it down. Quick as a wink, Erh-Lang stuffed the sun into the mountain peak and closed the top. The sun struggled but could not get out. The fate of the next five suns was the same.

The seventh sun was clever and observed what had happened to the others. The sun refused to rise in the sky. The sky faded to darkness and the Earth grew cold. "This will show the people," thought the proud sun.

The birds were the first to be afraid. "What will we do without any sun? The sky is too black for us to fly." The people called to the sun to come out, but the sun stayed hidden out of spite.

"Kind birds," pleaded the wise woman, "sing gently to the sun to let him know we will not hurt him. Please let him know how beautiful he is."

The birds began to sing sweetly while slowly, slowly the sun rose in the east. Erh-Lang and all the people cheered. The sun went proudly across the sky all day. In the evening, the sun rested.

The giant, Erh-Lang, is long gone from China. However, to this day, the birds sing every morning to coax the sun to rise in the sky.

Erh-Lang and the Seven Suns

Directions Using what you have just read, answer the questions.

1. Who ruled the skies in ancient China?

2. Why do you think that the people asked Erh-Lang for help?

3. How did the old woman in the story prove to be wise?

4. What happened after Erh-Lang put six suns in the mountain?

5. In the story, what does **spite** mean?

6. What might have happened if the suns continued their paths across the sky? Explain your answer.

Interrogative Pronouns

REMEMBER

An **interrogative pronoun** is used to introduce a sentence that asks a question. The interrogative pronouns are *who*, *whom*, *which*, and *what*.
Who, *whom*, and *whose* are used when speaking of persons.
Which is used when speaking of persons or things.
What is used when speaking of things, or when asking for information.
Some people confuse the use of *who* and *whom*. *Who* is the subject of a sentence and *whom* is used as the object of either a verb or a preposition.

EXAMPLES

Who can name the nine planets?

Whom did the teacher ask for help?

Whose answer is correct?

Which is closest to the sun?

What do you know about the stars?

Directions **Complete each sentence using one of these interrogative pronouns.**

who   whom   whose   which   what

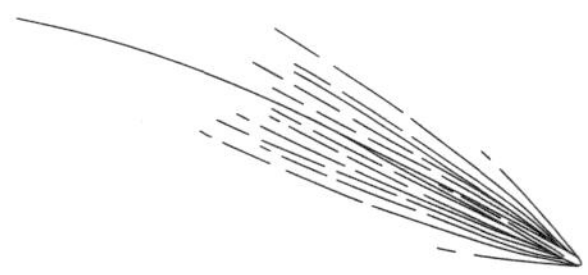

1. ______________________ discovered that the Earth revolves around the sun?
2. ______________________ are meteoroids made of?
3. ______________________ can describe a comet?
4. ______________________ of the planets is the smallest?
5. ______________________ discovered the planet Uranus?
6. ______________________ is the size of Saturn?
7. ______________________ telescope are you using?
8. For ______________________ is this comet named?
9. ______________________ of the planets is the coldest?
10. ______________________ does a radio telescope do?

Astronomy Analogies

Directions **Analogies are made from pairs of words that have the same relationship. Complete the analogies using words from the box below.**

Astronomy	constellation	galaxy	gravity	planet
revolve	satellites	Mars	telescope	universe

1. EARTH is to ______________________ as SUN is to STAR.
2. TEXAS is to STATE as ______________________ is to PLANET.
3. CLOSE is to MICROSCOPE as DISTANT is to ______________________.

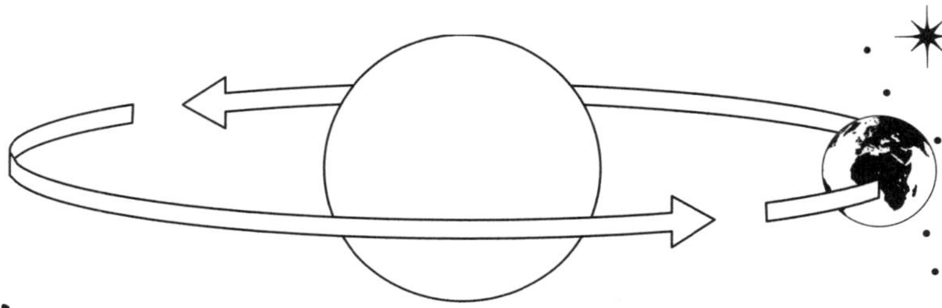

In the Night Sky

Directions **Read the essay. Use words from the box above to fill in the blanks.**

(4) ______________________ is the scientific study of the universe and all it contains. Everywhere, moons and other kinds of **(5)** ______________________ orbit planets. The planets **(6)** ______________________, or move around stars, held in their paths by the force of **(7)** ______________________. Billions of stars make up a **(8)** ______________________. Some of these stars form a pattern, such as the Big Dipper, which is called a **(9)** ______________________. The universe is so vast that there is much more out there beyond our **(10)** ______________________ Earth!

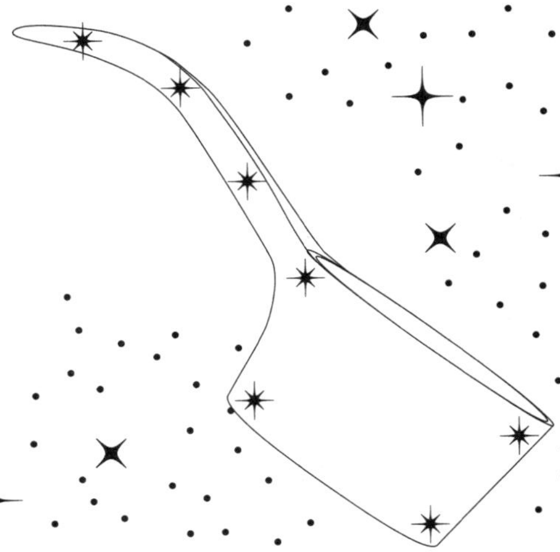

Multiplying Fractions

REMEMBER

When you multiply two fractions, you are finding part of a part.

To multiply two fractions:

- Multiply numerators.
  (This becomes the numerator of the answer.)
- Multiply denominators.
  (This becomes the denominator of the answer.)
- Then reduce the fraction to lowest terms.

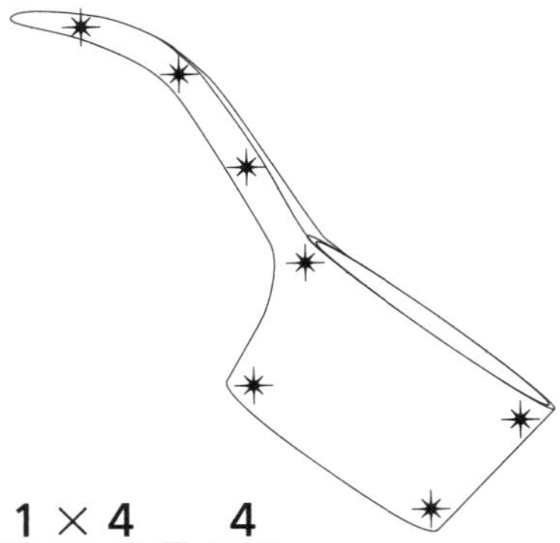

EXAMPLES

$$\frac{1}{2} \times \frac{1}{3} = \frac{1 \times 1}{2 \times 3} = \frac{1}{6} \qquad \frac{1}{3} \times \frac{4}{5} = \frac{1 \times 4}{3 \times 5} = \frac{4}{15}$$

$$\frac{3}{4} \times \frac{2}{3} = \frac{3 \times 2}{4 \times 3} = \frac{6}{12} = \frac{1}{2}$$

Fraction Action

Directions **Complete each multiplication problem. Reduce the answer to lowest terms. The first one is done for you.**

1. $\frac{1}{2} \times \frac{1}{2}$ can be written $\frac{1 \times 1}{2 \times 2} = \frac{1}{4}$

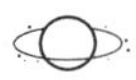

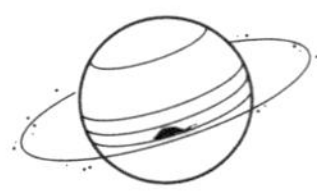

2. $\frac{1}{2} \times \frac{4}{5}$ can be written $\frac{\quad\times\quad}{\quad\times\quad}$ = ___ = ___

3. $\frac{1}{3} \times \frac{3}{5}$ can be written $\frac{\quad\times\quad}{\quad\times\quad}$ = ___ = ___

4. $\frac{2}{3} \times \frac{1}{2}$ can be written $\frac{\quad\times\quad}{\quad\times\quad}$ = ___ = ___

5. $\frac{3}{4} \times \frac{2}{5}$ can be written $\frac{\quad\times\quad}{\quad\times\quad}$ = ___ = ___

6. $\frac{5}{8} \times \frac{4}{6}$ can be written $\frac{\quad\times\quad}{\quad\times\quad}$ = ___ = ___

Fritter Fractions

Directions **Mani wants to make mashed potatoes, but he isn't sure he has enough ingredients. Use the information below to help Mani get started.**

Out-of-This-World Mashed Potatoes
4 servings

$1\frac{3}{8}$ lbs. potatoes

$\frac{1}{4}$ cup Parmesan cheese

$\frac{1}{3}$ cup vegetable broth

$\frac{3}{4}$ teaspoon salt

1. Mani has $\frac{3}{4}$ lb. Russet potatoes and $\frac{1}{3}$ lb. Idaho potatoes. Does he have enough potatoes to make 4 servings of his favorite recipe? Explain your answer.

____

____

2. After Mani harvests his potato patch, he plans to make $3\frac{1}{2}$ times more mashed potatoes than the original recipe.

a. How many servings does Mani plan to make? __________

b. How much of each ingredient will Mani need?

potatoes __________ Parmesan cheese ________

vegetable broth ________ salt __________

3. Later in the year, Mani is given $7\frac{1}{2}$ lbs. of potatoes. Mani has plenty of vegetable broth, salt, and cheese.

a. If Mani wanted to make one serving, what is the weight of the potatoes he would need? __________

b. To the nearest whole number, what is the greatest number of servings of mashed potatoes Mani can make? __________

Searching the Solar System

Directions Find the words listed in the box. Then circle them in the puzzle. The words can be hidden across, down, or diagonally.

Sun	Pluto	Saturn	Earth	Neptune
Mars	Jupiter	Venus	Mercury	Uranus

M	I	M	U	I	P	R	L	V	M
C	E	B	A	E	Y	L	Q	S	U
D	S	R	T	R	Y	P	U	A	S
J	N	E	C	P	S	E	N	T	I
U	R	A	N	U	S	Y	E	U	O
P	T	R	I	A	R	S	P	R	V
I	U	T	H	X	W	Y	T	N	K
T	N	H	S	R	U	M	U	P	T
E	R	I	A	U	V	E	N	U	S
R	L	R	H	Y	N	H	E	E	T

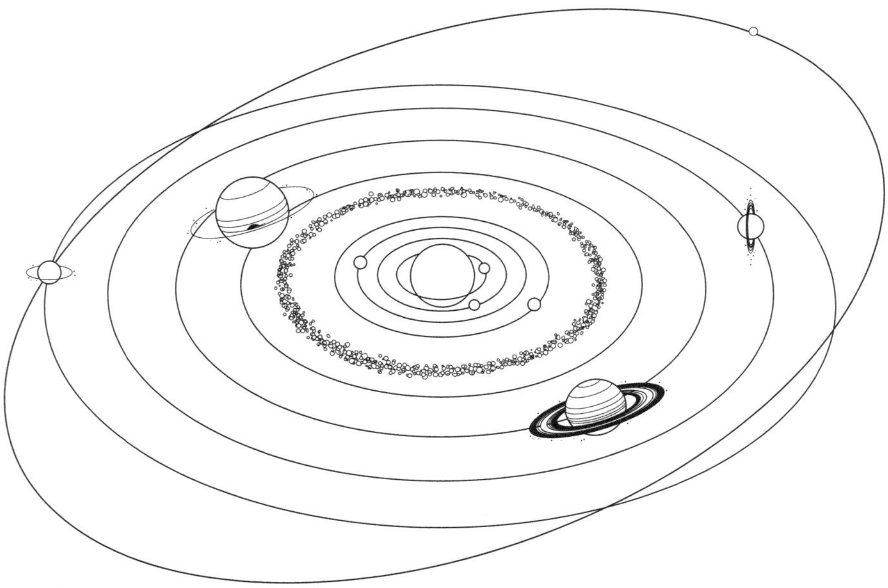

Alien Encounter

Directions Imagine you took a trip into space and met an alien. What would the alien's planet look like? What would the alien look like? Write a paragraph telling about your alien encounter. Don't forget to describe how the alien talks, what it eats, and how it dresses.

CHAPTER 4

The Wheel World

A young French inventor in the 18th Century had an idea that changed the way we travel. At that time, most people traveled by foot or on horseback. This inventor had an idea to combine the two to create a new way to travel. He worked on his idea until one day—voilà!—he appeared in the gardens of the Royal Palace with his new invention. It was made of wood and had two wheels. There was a crossbar for the rider to sit on and a carved horse's head to hold on to as he scooted along with his feet. People laughed, but other inventors soon started improving on this invention. The inventor didn't realize it at the time, but he had just created the first bicycle.

The bicycle has come a long way since it made its first appearance 200 years ago. Today, bicycles are better designed, more attractive, and easier to ride. No longer a curiosity, the bicycle has become a cheap and efficient way to travel. Cycling is a healthy form of recreation, too. For those who like to compete, it is also an exciting sport.

Sports cyclists have many choices, including road racing, track racing, and cross-country racing. In road racing, the riders race on the course to reach the finish line. The distances can be up to 150 miles or more. There are individual races, as well as team races. Some races, such as the Tour de France, last several weeks and cover great distances.

In track racing, cyclists race on indoor or outdoor oval tracks. Cross-country racing involves outdoor racing on terrain that is usually rugged and hilly. Off-road bicycling sports such as mountain biking and BMX (bicycle moto-cross) riding have captured the imagination of riders of all ages. These sports include tricks, as well as racing.

Each type of cycling, whether it's for travel, recreation, or sport, has its own special bike. There are even bicycles made just for doing tricks like riding on the back wheel or jumping over objects! People laughed at the world's first bicycle, but these days it's hard to imagine a world without bicycles.

The Wheel World

Directions Using what you have just read, answer the questions.

1. What did the first bicycle look like?

2. How do bicycles of today differ from when they were first invented?

3. Why do you think cycling is considered a healthy form of recreation?

4. What does the author mean by the phrase: *No longer a curiosity*?

5. What kind of bicycle would be best to use on rugged and hilly terrain?

6. What are the three ways bicycles are used?

Verbs: Action, Linking, and Helping

REMEMBER

An **action verb** is a word that expresses action, such as *jump, hike,* and *kick*.
The verb *be* is a **linking verb** and connects the subject of a sentence with a noun, pronoun, or adjective. Linking verbs tell what something is or is like.
Am, *is*, and *are* tell about things that are happening now.
Was and *were* tell about things that happened in the past.

A **verb phrase** is a group of words that does the work of a single verb.
The **main verb** names the main action in a verb phrase.
A **helping verb** comes before the main verb to help show action.

Common Helping Verbs

am	was	did	had	may	could
is	were	have	shall	can	should
are	do	has	will	might	would

EXAMPLES

Action verb: Orville and Wilbur Wright *built* a wind tunnel.
Main verb and helping verb: They *were exploring* a new frontier.
Linking verb: Their work *was* difficult.

Directions **Complete each sentence by adding an action verb, a helping verb, or a linking verb.**

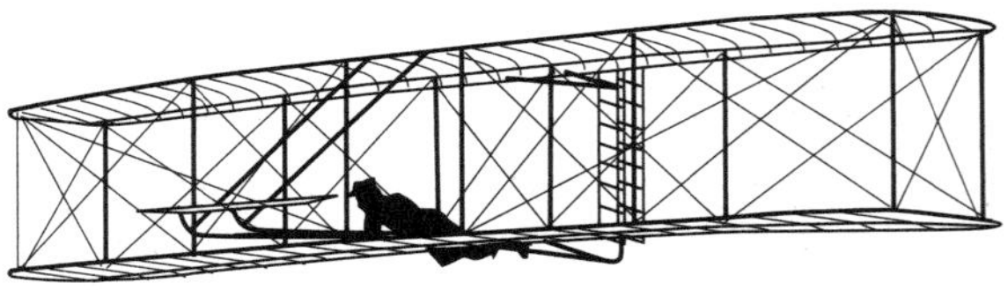

1. Orville ____________________ Wilbur about his new idea.
2. Wilbur ____________________ an idea of his own.
3. The brothers ____________________ many birds in flight.
4. They ____________________ many kites, too.
5. The Wright brothers ____________________ excited about their discoveries.
6. They ____________________ others to watch their test flights.

On the Move

Directions **Complete the analogies using words from the box below.**

bicycle	ferry	motorcycle	Taxicab
airport	sleigh	spacecraft	subway

1. THREE is to TRICYCLE as TWO is to ________________________.
2. GOAT is to BOAT as PLAY is to ________________________.
3. ROAD is to JEEP as WATER is to ________________________.
4. PARKING LOT is to CAR as ________________________ is to AIRPLANE.

To the Moon

Directions **Read the travel schedule. Use words from the box above to fill in the blanks.**

Travel Schedule

5:30 A.M. Put on helmet, get keys, drive **(5)** ________________________ to the "Yellow **(6)** ________________________ Company."

6:00 A.M. Take taxicab to the **(7)** ________________________ dock by the water.

6:20 A.M. Ferry departs.

6:45 A.M. Ferry reaches the mainland.

7:00 A.M. Walk to the underground **(8)** ________________________ station and take the subway to the **(9)** ________________________ to catch your flight.

7:20 A.M. Plane departs.

8:00 A.M. Arrive at the space station.

8:30 A.M. Board the **(10)** ________________________.

9:00 A.M. Rocket departs.

11:00 A.M. Arrive on the moon.

Dividing Fractions

REMEMBER

Follow these steps to divide a fraction by a whole number:

- Write the reciprocal of a whole number. (The reciprocal of a whole number is a fraction with 1 as the numerator and the whole number as the denominator.)
- Multiply the fraction by the reciprocal of the whole number.
- Then reduce to lowest terms.

EXAMPLES

$$\frac{1}{2} \div 4 = \frac{1}{2} \times \frac{1}{4} = \frac{1}{8}$$

$$\frac{2}{3} \div 4 = \frac{2}{3} \times \frac{1}{4} = \frac{2 \times 1}{3 \times 4} = \frac{2}{12} = \frac{1}{6}$$

The Great Divide

Directions **Write the reciprocal of each whole number for questions 1–4. Then complete questions 5–10.**

1. 4 ______ **2.** 3 ______ **3.** 2 ______ **4.** 5 ______

5. $\frac{1}{2} \div 2 = \frac{1}{2} \times \frac{1}{2} =$ ______

6. $\frac{1}{4} \div 3 =$ ______ $\times$ ______ $=$ ______

7. $\frac{2}{3} \div 4 =$ ______ $\times$ ______ $=$ ______

8. $\frac{3}{8} \div 3 =$ ______ $\times$ ______ $=$ ______

9. $\frac{8}{12} \div 2 =$ ______ $\times$ ______ $=$ ______

10. $\frac{9}{12} \div 3 =$ ______ $\times$ ______ $=$ ______

Fill It Up!

Directions **The gas gauge on Paula's car is shown below. The gas tank holds 16 gallons when full. Paula goes to a gas station to fill the tank. Use this information to complete the page.**

1. What fraction of Paula's gas tank is full? ________

2. How many gallons of gas does Paula have left in the tank? ________

3. What fraction of Paula's gas tank is empty? ________

4. How many gallons of gas will it take to fill the gas tank? ________

5. Three days later, Paula's gas tank is $\frac{3}{4}$ full. She is not sure if she should stop for gas.

a. What fraction of the tank is empty? ________

b. How many gallons will it take to fill the tank? ________

c. Should Paula stop for gas? Explain your answer.

____

____

Ride Riddles

Directions **Use the clues to solve the riddles. Hint: Think about words that rhyme with those in dark print.**

EXAMPLE: What kind of vehicles do they drive on **Mars**?

M a r s   c a r s

1. What do you call a **truck** that cannot get out of the mud?

A ___ ___ ___ ___ ___   t r u c k

2. What do you call a man who pretends to be a **tractor**?

A t r a c t o r   ___ ___ ___ ___ ___

3. What do you call an undecorated **plane**?

A ___ ___ ___ ___ ___   p l a n e

4. What do you call the effort a **train** uses to struggle up a mountain?

A t r a i n   ___ ___ ___ ___ ___ ___

5. What do you call the **trip** a big boat would take?

A ___ ___ ___ ___   t r i p

6. What do you call a **new** boat?

A n e w   ___ ___ ___ ___ ___

Cars of the Future

Directions Cars have changed since they were first invented nearly one-hundred years ago. As our world changes, how will it affect cars? What do you think they will look like a hundred years from now? Draw a picture of what the car of the future will look like. Then write a paragraph describing the car.

CHAPTER 5

Leonardo da Vinci

The Renaissance, a time of artistic revival in Europe, began in Italy in the 14th Century. The Renaissance spread throughout Europe and lasted for about 200 years. During this time, people were interested in art, literature, and learning. Some of the buildings constructed during the Renaissance are still standing today. There are also many paintings, statues, and other pieces of artwork that we still enjoy.

One of the many artists to become famous during this time was Leonardo da Vinci. Leonardo was born in 1425 in Vinci, Italy. He spent his childhood drawing everything he saw. As a little boy he drew detailed and beautiful pictures. His father thought it was important for him to learn more about art and drawing. When Leonardo was a teenager, his father arranged for him to move to Florence, Italy, to work with the famous painter, Andrea del Verrocchio. When Leonardo was 20 years old, Andrea asked him to help finish a painting. When people saw the painting, many of them thought the angel Leonardo painted was the best part. After that, Leonardo completed paintings on his own.

Art and science always interested Leonardo. He studied the human body and many things in nature. Many people believe his love of science made his paintings more realistic. He was famous for the pictures he drew and painted of people, too. His most famous painting of a woman, titled the "Mona Lisa," was painted from 1503–1506. This painting is well-known for Mona Lisa's eerie smile. No matter where you stand to look at the painting, Mona Lisa looks straight at you. This painting hangs in the Louvre Museum in Paris, France. Most of Leonardo's works hang in European museums.

When Leonardo was not painting, he was exploring architecture, engineering, and science. Flight fascinated him. He spent hours watching birds fly and taking very detailed notes. He wrote these notes backwards. To read them, the notes must be held up to a mirror.

Drawings and notes showing sketches of machines that look similar to present-day helicopters and gliders fill his notebooks. He even built some of these things. Since engines were not yet invented, Leondardo thought man's best chance to fly was with a glider. His notes contain many other drawings and descriptions of inventions he was not able to make. Leonardo lived about 500 years ago, so he did not have many of the tools or materials needed to create some of the things he invented. There were no electric lights, cars, or planes at that time. Materials such as plastic and rubber were not available.

Scientists have studied Leonardo's inventions and experiments for hundreds of years. Many of his ideas have been proven to work now that more advanced materials and tools are available. Imagine having ideas and inventions that are hundreds of years ahead of their time!

Leonardo da Vinci

Directions **Using what you have just read, answer the questions.**

1. Why did Leonardo's interest in science make him a better artist?

____

____

2. How do you think Andrea del Verrocchio felt when people liked Leonardo's work more than his?

____

____

3. Was it a good idea for Leonardo's father to send him to Florence? Why or why not?

____

____

4. Why didn't Leonardo make all the machines he drew?

____

____

5. Do you think Leonardo's life would have been different if he were born the same year as you? How?

____

____

6. Why did the author write this article?

____

____

Adjectives and Adverbs

REMEMBER

An **adjective** describes a noun or a pronoun. Adjectives usually come before the nouns they describe. An adjective can come after a linking verb and describe the subject of a sentence.
The words *a*, *an*, and *the* form a special group of adjectives called articles. Some adjectives, such as *this*, *these*, *that*, and *those*, point out a definite person, place, or thing. Other adjectives tell how many.

An **adverb** describes a verb, adjective, or another adverb. Adverbs tell *when*, *where*, or *how*. Adverbs also compare two or more actions.

EXAMPLES

Adjective: We entered *the* museum through *two huge*, *wooden* doors.
Adverb: We speak *softly* in the museum.

Directions **Read each sentence. Write if the underlined word is an adjective or an adverb on the line.**

1. We were curious to see the museum's collection. ________________
2. The room contained Japanese paintings. ________________
3. I walked carefully among the sculptures. ________________
4. The museum had fourteen guides. ________________
5. All the guides are friendly. ________________
6. Will you return to the museum tomorrow? ________________
7. This hall features modern paintings. ________________
8. The tour was over quickly. ________________

Art Words

Directions **Choose a word from the box below that matches each definition. Write it on the line.**

canvas	charcoal	collage	exhibit	Gallery
original	portrait	scarlet	sculpture	sketch

1. A picture made by cutting and gluing things onto paper ____________________

2. Dark gray ____________________

3. An innovative or creative work ____________________

4. A piece of framed cloth you paint on ____________________

5. Bright red ____________________

6. A rough drawing ____________________

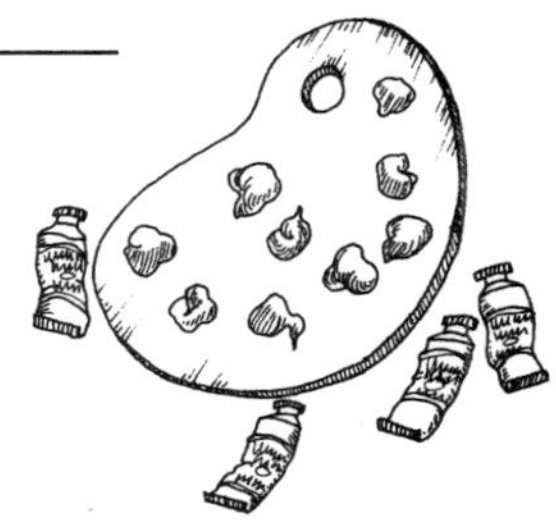

Art in Review

Directions **Read the art review. Use words from the box above to fill in the blanks.**

Last week's art **(7)** ____________________ at the newly renovated Bella Belling Art **(8)** ____________________ included many new and **(9)** ____________________ works of art. We saw a self- **(10)** ____________________ that gave a striking image of the artist as a young woman. There was also an unusual landscape painted on **(11)** ____________________. I especially liked the fantastic bronze **(12)** ____________________ that took up much of the lobby!

Ratios

REMEMBER

A **ratio** is a comparison of two numbers. A ratio is often written as a fraction.

EXAMPLES

The ratio of paintings to sculptures in a museum is 27 to 15. As a fraction, the ratio is written $\frac{27}{15}$.

The ratio of sculptures to paintings in a museum is 15 to 27. As a fraction, the ratio is written $\frac{15}{27}$.

To simplify a ratio:

- Reduce to lowest terms.

  Example $\frac{6}{9} = \frac{6 \div 3}{9 \div 3} = \frac{2}{3}$

- Do not write an improper fraction ratio as a mixed number.

  Example $\frac{4}{3}$ is NOT reduced to $1\frac{1}{3}$.

- Write a whole number ratio with a denominator.

  Example 8 to 2 is $\frac{8}{2}$ or $\frac{4}{1}$.

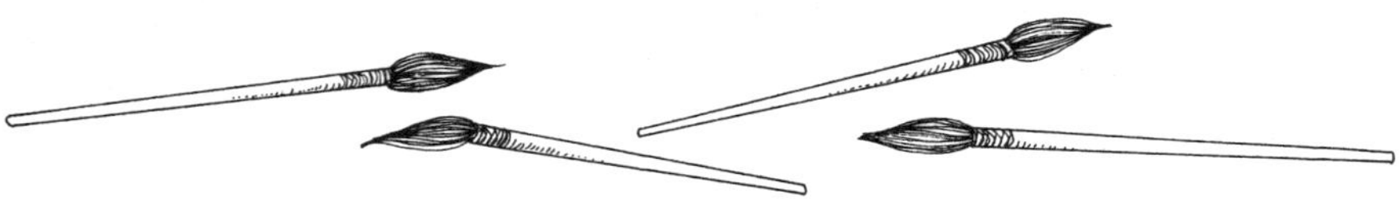

Ratio to Fraction

Directions **Write each ratio as a fraction. Then reduce to lowest terms.**

1. 4 to 8 ______ = ______ **2.** 6 to 10 ______ = ______ **3.** 15 to 20 ______ = ______

4. 8 to 6 ______ = ______ **5.** 10 to 6 ______ = ______ **6.** 20 to 5 ______ = ______

7. 12 to 4 ______ = ______ **8.** 8 to 4 ______ = ______ **9.** 50 to 10 ______ = ______

10. 80 to 4 ______ = ______ **11.** 18 to 16 ______ = ______ **12.** 81 to 9 ______ = ______

Bowled Over

Directions Use the information below to answer the questions.

Olivia made several bowls in art class. Each bowl weighs $\frac{3}{4}$ lb. The total weight of clay Olivia used for the bowls was $6\frac{3}{4}$ lbs. How many bowls did Olivia make?

1. Assume that n stands for the number of bowls that Olivia made.

a. Circle the equation that is true for n.

$6\frac{3}{4} \div n = \frac{3}{4}$ $\quad$ $6\frac{3}{4} \times \frac{3}{4} = n$ $\quad$ $6\frac{3}{4} - \frac{3}{4} = n$

b. How many bowls did Olivia make? ________

Olivia made 6 plates in art class. Each plate weighed $1\frac{1}{2}$ lbs. How many pounds of clay did Olivia use for these plates?

2. Assume that p stands for the pounds of clay that Olivia used.

a. Circle the equation that is true for p.

$p \div 1\frac{1}{2} = 6$ $\quad$ $p \times 1\frac{1}{2} = 6$ $\quad$ $6 \div \frac{1}{12} = p$

b. How many pounds of clay did Olivia use for the plates? ________

Olivia used $6\frac{7}{8}$ lbs. of clay to make 11 teacups. How much clay did she use for each teacup?

3. Assume that c stands for the amount of clay in each teacup.

a. Write an equation that is true for c. ____________________

b. How many pounds of clay did Olivia use for each teacup? ________

Masterpiece Mystery

Directions Look at the code. Each number stands for a letter. Write the letters on the lines to solve the riddles.

A	B	C	D	E	F	G	H	I	J
1	2	3	4	5	6	7	8	9	10
K	L	M	N	O	P	Q	R	S	T
11	12	13	14	15	16	17	18	19	20
U	V	W	X	Y	Z				
21	22	23	24	25	26				

1. Why did the artist take golf lessons?

___ ___ ___   ___ ___ ___ ___ ___ ___   ___ ___   ___ ___ ___ ___ ___
19 8 5   23 1 14 20 5 4   20 15   2 18 21 19 8

___ ___   ___ ___   ___ ___ ___   ___ ___ ___ ___ ___ ___ ___.
21 16   15 14   8 5 18   19 20 18 15 11 5 19

2. Why couldn't the detective solve the case of the missing charcoal drawing?

___ ___ ___   ___ ___ ___ ___ ___   ___ ___ ___ ___
20 8 5   3 12 21 5 19   23 5 18 5

___ ___ ___ ___ ___ ___ ___.
19 11 5 20 3 8 25

3. What did the art student say when his self-portrait won first place?

___   ___ ___   ___ ___ ___ ___ ___ ___ ___
9   1 13   16 9 3 20 21 18 5

___ ___ ___ ___ ___ ___ ___!
16 5 18 6 5 3 20

4. Why was the cowboy such a good artist?

___ ___   ___ ___ ___   ___ ___ ___ ___ ___   ___ ___
8 5   23 1 19   17 21 9 3 11   15 14

___ ___ ___   ___ ___ ___ ___.
20 8 5   4 18 1 23

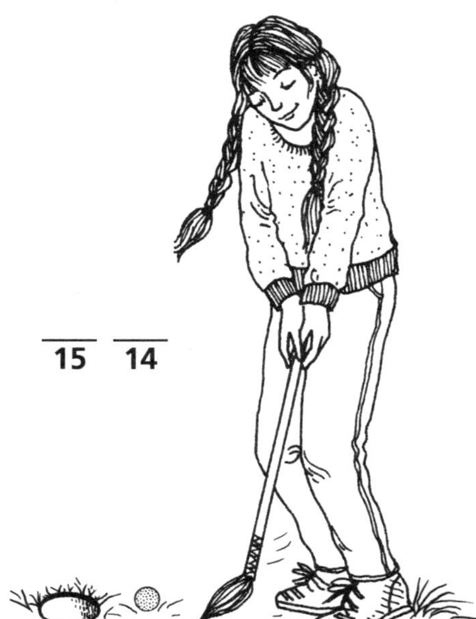

Self-Portrait

Directions A self-portrait often shows an image of the artist. It could show what the artist looks like or feels like. Draw your self-portrait showing what you look or feel like. Then write a few sentences about your portrait.

CHAPTER 6

Postcards from Rome

May 4

Dear Kuo,

I just arrived for my two-week stay in Rome. The flight was pleasant, but the car ride was much more interesting. Crisscrossing the countryside around Rome are large stone arches that connect together, forming narrow concrete paths. They almost look like bridges, but they are much too narrow. I am fascinated by this ancient city! I hope that during my stay I can unravel some of the mysteries I see around me.

Your friend,

Maddy

May 7

Dear Kuo,

I learned that those odd sets of arches are called aqueducts (AK-wuh-duhkts). Aqueducts were an ancient plumbing system that carried water from the mountains and hills surrounding Rome into the city. The water flowed through tunnels and into a channel at the top of the concrete and stone arches. These aqueducts were built between 312 B.C. and A.D. 455. People could use a public fountain to fill up a jar—rather than walking all the way to a mountain spring. The wealthiest people had water piped directly into their homes. They had running water when they needed it simply by turning on a faucet. Just imagine, two thousand years ago people had running water in their homes, just like a modern kitchen. The ancient Romans were incredible plumbers!

Your friend,

Maddy

Postcards from Rome

Directions Using what you have just read, answer the questions.

1. What is this selection mostly about?

2. What are **aqueducts**?

3. Aqueducts were built to slope downhill. Why do you think the Romans built them like this?

4. What does the word **unravel** in this section mean?

5. How were some homes in ancient Rome like those we live in today?

6. Find a sentence in the first postcard that is an opinion and write it on the lines.

Prepositions

REMEMBER

A **preposition** is a word that shows the relationship between a noun or pronoun and another word in a sentence. Some common prepositions are *about, across, after, before, beneath, between, from, into, near, of, on, past, since, up, with,* and *without.*

A **prepositional phrase** is a group of words that begins with a preposition and ends with a noun or pronoun. You already know that an adjective describes a noun or a pronoun. A prepositional phrase can act as an adjective.

EXAMPLES

Melissa sat *on the porch* every evening.
The poles *for the tent* were missing.

Camp Arrowhead

Directions **Draw one line under each prepositional phrase used as an adjective. Draw two lines under the noun or pronoun it describes.**

1. The counselors at the camp are friendly.
2. Penny is the name of our counselor.
3. She tells funny stories about Camp Arrowhead.
4. I took a picture of Penny.
5. The four girls near the cabin are sisters.
6. Two of them are twins.
7. Girls from many states come here.
8. The food at the camp is delicious.
9. Gardens near the camp supply fresh vegetables.
10. The river below the camp is beautiful.

Travel Analogies

Directions Complete the analogies using words from the box below.

excursion	gasoline	gear	exciting	luggage
motel	takeoff	tourist	transportation	vacation

1. TOP is to BOTTOM as ____________________ is to LANDING.

2. ESCAPIST is to ESCAPE as ____________________ is to TOUR.

3. DOG is to ANIMAL as TRAIN is to ____________________.

4. JACKET is to COAT as TRIP is to ____________________.

A Quick Break

Directions Read the letter. Use words from the box above to fill in the blanks.

Dear Chandra,

My mom and I are getting ready for our big week-long **(5)** ____________________ out West. It's going to be an **(6)** ____________________ trip. We'll put our **(7)** ____________________ in the trunk, fill up the car's tank with **(8)** ____________________, put the car into **(9)** ____________________ and go! We plan to stop at a **(10)** ____________________ along the way to sleep. That's where we'll scout out all the best **(11)** ____________________ attractions. I can't wait! I'll write more at our next stop.

Love, Thalia

Decimals

REMEMBER

Decimals, like fractions, represent a whole. The first decimal place is the **tenths** place. The second decimal place is the **hundredths** place. The third decimal place is the **thousandths** place, or 1000^{ths} place.

EXAMPLES

Decimal	In words	Meaning
0.3	three tenths	3 parts out of 10
0.03	three hundredths	3 parts out of 100
0.003	three thousandths	3 parts out of 1000

To the Point

Directions **Write each number using decimals.**

1. five thousandths ______________

2. $\frac{6}{100}$ ______________

3. four hundredths ______________

4. $\frac{78}{1000}$ ______________

5. one hundred nine thousandths ______________

6. $\frac{253}{1000}$ ______________

Directions **Write each number using words.**

7. 0.006 ______________________________

8. 0.7 ______________________________

9. 0.13 ______________________________

10. 0.377 ______________________________

On Your Mark

Directions **Use the information on the chart to complete the page.**

Each year, Lincoln Middle School holds the Lincoln Spring Track Meet. The five best times for boys in the 400-meter run are shown in the chart. Each time is given to the nearest thousandth of a second.

Lincoln Spring Track Meet

Name	Time (in seconds)
Darrell	69.750
Mickey	68.467
Gunnar	69.073
Nick	69.942
Gerald	68.875

1. Write the runners' names from fastest to slowest.

1st: ______________________ (fastest time)

2nd: ______________________

3rd: ______________________

4th: ______________________

5th: ______________________ (slowest time)

2. By how many seconds did the fastest runner finish ahead of the slowest runner?

____

3. Which two runners finished closest to the same time?

______________________ ______________________

4. By how many seconds did Mickey miss tying the meet record of 67.954 seconds?

____

5. What was the median score of the runners? ______________________

6. What was the average of the five race times? Round to the nearest thousandth.

____

Travel Twisters

Directions Read each clue. Unscramble the letters to find the words. Write the words on the lines. The use the numbered letters to solve the riddle.

1. wbsayu An underground train system

___ ___ ___ ___ ___ ___
(letter 4 = 1, letter 5 = 2)

2. stardcop A card you send someone when you are on a trip

___ ___ ___ ___ ___ ___ ___ ___
(letter 3 = 3)

3. glagueg The bags a person carries on a trip

___ ___ ___ ___ ___ ___ ___
(letter 1 = 4, letter 2 = 5)

4. praitro The place you go to catch a flight on a plane

___ ___ ___ ___ ___ ___ ___
(letter 2 = 6)

5. feotkaf To climb or jump into the air on a plane

___ ___ ___ ___ ___ ___ ___
(letter 1 = 7)

6. kcpa To fill a suitcase for a trip

___ ___ ___ ___
(letter 3 = 8)

7. vartel To go from place to place

___ ___ ___ ___ ___ ___
(letter 5 = 9)

NOW SOLVE THE RIDDLE: What kind of luggage does a lawyer carry?

___   ___ ___ ___ ___ ___ ___ ___   ___ ___ ___ ___

2   4 2 1 3 5 6 7   8 2 3 9

Post This!

Directions Imagine you're planning a cross-country car trip. What places would you like to visit? What would you see there? Send your friend a postcard from one of your destinations. Draw a picture of your destination on the front of the postcard and write your message on the back.

(front)

To: ____________________

____

____

____

(back)

CHAPTER 7

Strong Wind

In old times, a great Indian warrior had a strange and wonderful power: he could make himself invisible. He was known as Strong Wind, the Invisible. He lived with his sister and together they did great deeds.

Many maidens wanted to marry Strong Wind. It was known that Strong Wind would marry the first maiden who could see him when he came home each evening.

Strong Wind used a clever trick to test the truthfulness of all who sought to marry him. Each evening, his sister walked on the beach with a girl wishing to marry Strong Wind. As Strong Wind approached, his sister would ask the girl, "Do you see him?" Each girl would falsely answer, "Yes." Strong Wind's sister would then ask, "With what does he pull his sled?" The girls would reply, "with the hide of a moose," or "with a pole." The sister would know that they had lied and their answers were mere guesses.

A great chief and his three daughters lived in the village. The youngest daughter, Glenna was beautiful and gentle. Her older sisters were very jealous of her and treated her cruelly. They made her dress in rags and cut off her beautiful, long, black hair. Glenna remained patient despite this and went gladly about her work.

Like other girls, the chief's two eldest daughters wanted to marry Strong Wind. One evening they walked with Strong Wind's sister. When he came home from his day's work, drawing his sled, Strong Wind's sister asked, "Do you see him?"

Each sister lied, answering, "Yes."

She asked, "Of what is his shoulder strap made?"

Each one, guessing, said, "Of rawhide." Strong Wind's sister knew they had lied and sent them away.

One day, Glenna resolved to see Strong Wind. She patched her clothes, put on the few ornaments she possessed, and went forth to see Strong Wind. Her sisters laughed at her, but she silently went on her way.

Strong Wind's sister received the girl kindly and took her to the beach. Soon Strong Wind came home pulling his sled. His sister asked, "Do you see him?"

"No," answered Glenna, speaking the truth. Again Strong Wind's sister asked, "Do you see him?"

Glenna answered, "Yes, and he is very wonderful."

"With what does he pull his sled?" asked Strong Wind's sister.

"With the rainbow," Glenna answered. Strong Wind's sister knew that because Glenna had spoken the truth, Strong Wind had made himself visible to her. Strong Wind's sister gave Glenna fine clothes to wear and many rich ornaments. When she brushed Glenna's hair, it grew back as long and beautiful as it once was. The very next day, Glenna and Strong Wind married and together they did great deeds.

Strong Wind

Directions **Using what you have just read, answer the questions.**

1. What power did Strong Wind possess?

2. In what two ways did Strong Wind's sister help him?

3. Why did the chief's older daughters treat Glenna badly?

4. How did Glenna prepare herself to go to see Strong Wind?

5. Why did Strong Wind show himself to Glenna?

6. How is this story like the fairytale *Cinderella*?

Interjections

REMEMBER

An **interjection** is a word or group of words that expresses a strong or sudden feeling. An interjection that expresses a mild feeling is followed by a comma. An interjection that expresses a strong feeling is usually set off from the rest of the sentence by an exclamation point. An interjection can be part of a sentence, or it can stand alone. It is not grammatically connected to any other word in the sentence.

Common Interjections

Ah	Gee	Hooray	Oh, no
Aha	Goodness	Hush	Oops
Beware	Great	Oh	Shh
Bravo	Hey	Oh, my	Wow

EXAMPLES

Wow, this is the biggest swimming pool I've ever seen!
Ouch! The pavement is hot!

Express Yourself!

Directions **Write the interjection on the line next to each sentence.**

1. Wow! This pool is as long as a football field! ______________________
2. Gee, look at all the people here. ______________________
3. Hey, let's jump in the water! ______________________
4. Yikes! The water is cold! ______________________
5. Oops, I forgot my goggles. ______________________
6. Shh, do you hear that? ______________________
7. Oh, no! That was a loud clap of thunder. ______________________
8. Goodness, that's a dark cloud coming our way! ______________________
9. Oh, my! We have to get out now. ______________________
10. Great! Now it's going to rain. ______________________

Climate Comparisons

Directions Complete the analogies using words from the box below.

blizzard	weather	drought	tornado
hurricane	lightning	sandstorms	thunderstorms

1. RAIN is to HURRICANE as SNOW is to ______________________.

2. CRASH is to THUNDER as FLASH is to ______________________.

3. RAY is to SUNLIGHT as FUNNEL is to ______________________.

4. FLOOR is to CEILING as FLOOD is to ______________________.

Weather Watch

Directions Read the weather report. Use words from the box above to fill in the blanks.

It is snowing so hard, they're calling it a **(5)** ______________________ in the Northern region of the country. If this were summertime, we could expect loud **(6)** ______________________ with **(7)** ______________________ to light up the sky. Other parts of the country will experience strong gusts of wind. They won't be strong enough to classify as a **(8)** ______________________, though. There's also a chance of **(9)** ______________________ in the desert. They've had enough rain there this year to end the **(10)** ______________________! It really has been an interesting day across the country. The **(11)** ______________________ is different everywhere!

Multiplying Decimals

REMEMBER

To multiply a decimal by a whole number:

- Multiply the numbers.
- Then place a decimal point in the answer. Use the same number of decimal places as in the number being multiplied.

EXAMPLES

$$\begin{array}{r} 5.3 \\ \times\ 4 \\ \hline 21.2 \end{array}$$

5.3 ← 1 decimal place; 21.2 ← 1 decimal place

$$\begin{array}{r} 3.26 \\ \times\ 7 \\ \hline 22.82 \end{array}$$

3.26 ← 2 decimal places; 22.82 ← 2 decimal places

Decimal Times

Directions Multiply to find the product.

1. $\begin{array}{r} 0.3 \\ \times\ 4 \\ \hline \end{array}$

2. $\begin{array}{r} 0.2 \\ \times\ 3 \\ \hline \end{array}$

3. $\begin{array}{r} 0.1 \\ \times\ 8 \\ \hline \end{array}$

4. $\begin{array}{r} 2.9 \\ \times\ 4 \\ \hline \end{array}$

5. $\begin{array}{r} 3.7 \\ \times\ 5 \\ \hline \end{array}$

6. $\begin{array}{r} 8.15 \\ \times\ 2 \\ \hline \end{array}$

7. $\begin{array}{r} \$0.02 \\ \times\ 2 \\ \hline \end{array}$

8. $\begin{array}{r} \$0.72 \\ \times\ 5 \\ \hline \end{array}$

9. $\begin{array}{r} \$1.33 \\ \times\ 4 \\ \hline \end{array}$

A Change in the Air

Directions **Follow the directions to complete the page.**

1. Write the value of each coin as a decimal and as a fraction.

Coin	Decimal	Fraction
a. penny	________	________
b. nickel	________	________
c. dime	________	________
d. quarter	________	________

2. Write the value of each group of coins as a decimal and as a fraction or mixed number.

	Decimal	Fraction
a.  1 dime, 2 nickels, 3 pennies	________	________
b.  1 dime, 3 nickels, 4 pennies	________	________
c.  One $1 bill and 2 quarters	________	________
d.  Three $1 bills, 2 dimes, 1 penny	________	________

Riddled with Weather

Directions Use the clues to solve the riddles. Hint: Think about words that rhyme with those in dark print.

EXAMPLE: What do you call a car being pulled out of the **snow**?

A s n o w t o w

1. What do you call a **pail** filled with small ice particles?

A ___ ___ ___ ___ p a i l

2. What do you call a question that is asked about **thunder**?

A t h u n d e r ___ ___ ___ ___ ___ ___

3. What would you call a dull, boring **rain**?

___ ___ ___ ___ ___ r a i n

4. What sound does **drizzle** make when it hits the hot street?

d r i z z l e ___ ___ ___ ___ ___ ___

5. What do you call several clouds in a large group?

A c l o u d ___ ___ ___ ___ ___

Weather People

Directions Many Native American tales use people to represent things that happen in nature, such as thunder or snow. Write a paragraph describing a person that would represent the rain. What would this person look like? What would he or she wear? Where would this person live? Draw a picture of this person.

CHAPTER 8

Before the QUAKES

In the middle of the night, a pager goes off and a scientist is awakened. He knows instantly what has happened. Somewhere in the world, a major earthquake has struck. The scientist is a seismologist (size-MAWL-uh-jist), someone who studies earthquakes. He works for the National Earthquake Information Center (NEIC), which locates major earthquakes that occur around the world.

All NEIC employees wear pagers so that they can be alerted when a major earthquake happens. They use computer information to determine exactly where the quake occurred and how powerful it was. NEIC employees then supply this information to emergency workers. They describe the earthquake to newspaper, radio, and television reporters, and tell them what can be expected in that area. Often aftershocks, or smaller earthquakes, can follow a larger quake.

How does a computer get information about an earthquake? It receives information directly from a seismograph (SIZE-muh-graf), a machine that records the motion of the ground during an earthquake. A seismograph receives signals from a seismometer (size-MOM-it-tur), an instrument that senses ground movement.

Can scientists predict earthquakes the way weather forecasters predict storms? It's difficult to make guesses about the forces of nature. The forces that control earthquakes are complicated. Scientists often know where an earthquake will likely happen, but they cannot tell when it will strike. Even if seismologists could predict earthquakes, they couldn't control them. Earthquakes happen at places underground where huge rocks on either side of a crack push against each other. The rocks snap past each other, which causes the ground above to shake and crack. The energy of the quake spreads outward in all directions and can cause great damage in places where people live.

Seismologists spend their time finding ways to reduce earthquake damage. Some work on building construction. They study problems that earthquakes could cause to buildings, dams, and bridges. Others help design structures that can withstand the shock of earthquakes. Even if they can't stop earthquakes, these scientists hope their work will help save lives when a major quake strikes.

EARTHQUAKE EARTHQUAKE EARTHQUAKE

Before the Quakes

Directions **Using what you have just read, answer the questions.**

1. What is this article mostly about?

2. Why is it important for a seismologist to have a pager?

3. What might happen if the seismologist did not have a pager when an earthquake occurred?

4. Why do news reporters need to be told about earthquakes?

5. What is a seismograph?

6. How could scientists working in construction help reduce earthquake damage?

Sentence Subjects and Predicates

REMEMBER

A sentence is a group of words that expresses a complete thought. Every sentence has two main parts: the **subject** and the **predicate**. The subject tells who or what the sentence is about. The predicate tells what the subject does or is like.

The **simple subject** is a noun or pronoun. It is the main word in the complete subject. The **complete subject** includes the simple subject and any words telling about the subject.

The **simple predicate** is the verb or verb phrase that expresses action or a state of being. The **complete predicate** includes the verb and any words that tell about the action or complete its meaning.

EXAMPLE

complete subject **complete predicate**

Foreign bugs are invading the United States.

simple predicate

simple subject

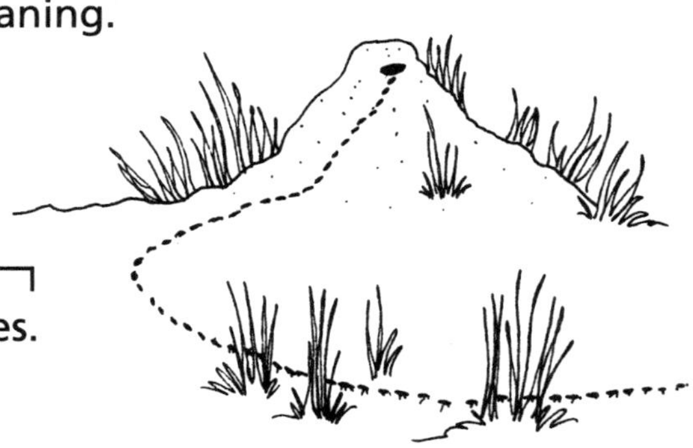

Directions **Draw a line between the complete subject and the complete predicate in each sentence. Then write each simple subject and each simple predicate on the lines.**

1. Many scientists are studying these bugs. ______________ ______________
2. The fire ant is from South America. ______________ ______________
3. This ant has a blistering sting. ______________ ______________
4. Fire ants have invaded the south. ______________ ______________
5. They are slowly creeping north. ______________ ______________
6. Their stings can be very painful. ______________ ______________
7. The bug feasts on native insects and destroys crops. ______________ ______________
8. The ant has few natural enemies in the United Sates. ______________ ______________

Scientific Skills

Directions **Choose a word from the box below that matches each definition. Write it on the line.**

biology	chemistry	ecology	environment	technology
evaporate	geology	pollute	research	

1. A science that studies the relationship between organisms and their environments ____________________
2. To collect information on a particular subject ____________________
3. To contaminate with waste or to make unclean ____________________
4. To change into a vapor ____________________

Modern Science

Directions **Read the article. Use words from the box above to fill in the blanks. You can use a dictionary for help.**

Scientists have been doing in-depth **(5)** ____________________ for many years. Advances in modern **(6)** ____________________ have helped them learn more about our world. It is possible to use this technology in ways that do not pollute the **(7)** ____________________ around us. Scientists are making advances in the study of life and living things, or **(8)** ____________________, as well as in **(9)** ____________________, which is the study of the properties of chemicals. Scientists studying the relationship between organisms and their environments, or **(10)** ____________________ have also learned many new things. They now know more about our world than ever before.

Dividing Decimals

REMEMBER

To divide a decimal by a whole number:

- Divide the numbers.
- Then place a decimal point in the quotient directly above the decimal point in the dividend.

EXAMPLES

0.62 4)2.48	5.01 5)25.05	$2.40 3)$7.20

Decimal Division

Directions Divide to find the quotient.

1. 3)0.6 **2.** 8)0.16 **3.** 3)0.09

4. 4)1.28 **5.** 7)2.17 **6.** 4)9.52

7. 2)$1.68 **8.** 5)$8.25 **9.** 6)$47.52

Value of Equations

An **equation** may contain more than one addition, subtraction, multiplication, or division step. Learning to find the value of an equation is an important skill.

Dalia and Bryon try to find the value of the equation: $5 \times 6 + 6.6 \div 3$.

Dalia's Solution
Step 1. Multiply: $5 \times 6 = 30$
Step 2. Add: $30 + 6.6 = 36.6$
Step 3. Divide: $36.6 \div 3 = 12.2$
Dalia's Answer: 12.2

Bryon's Solution
Step 1. Add: $6 + 6.6 = 12.6$
Step 2. Multiply: $5 \times 12.6 = 63$
Step 3. Divide: $63 \div 3 = 21$
Bryon's Answer: 21

Both of these answers are wrong! Do you know why?

Dalia and Bryon have not learned the rules for finding the value of an equation:

Follow the order of operations to simplify an expression.
The table below shows the order.

Order of Operations
• Do all operations within the parentheses first.
• Do all exponents before other operations.
• Multiply and divide in order from left to right.
• Add and subtract in order from left to right.

Directions Find the following values.

1. $(2.4 \times 3) + (3.2 \div 4) =$

2. $(9.6 \div 6) - (0.2 \times 3) =$

3. $(1.8 \times 2) + (15.5 \div 5) =$

4. $(21.6 \div 3) + (1.2 \times 4) =$

5. $(39.76 \div 7) - (2.34 \times 2) =$

Puzzling Science

Directions Use words from the box below to finish the puzzle. Use a dictionary for help, if needed.

archeology	biology	botany	chemistry	ecology
geology	meteorology	physics	seismology	zoology

ACROSS

1. The science of earthquakes
4. The study of chemicals and their properties
6. The study of rocks
7. The study of past human life
8. The study of living things
9. The study of plants
10. The study of matter and energy

DOWN

2. The study of weather
3. The study of living things in their environments
5. The study of animals

Dino-mite Discovery!

Directions Archeologists study the past. They get much of their information from fossils. Imagine that you are an archeologist and have found the bones of an undiscovered dinosaur. Draw a picture of what that dinosaur looked like. Name your discovery and write a few sentences about what the dinosaur ate, how it moved, and where it lived.

CHAPTER 9

America's Hometown

When you think of Washington, D.C., what comes to mind? You may think it is the center of government and politics, or the site of many national monuments. Washington, D.C., is this and much more! When you look beyond its importance as the nation's capital, you find an exciting, international city. You also see a city that, in some ways, is like other cities you may know.

Washington is on the east coast of the United States. It lies between the Potomac and Anacostia Rivers. The "D.C." stands for District of Columbia. A city without a state, Washington is located inside a federal district named after Christopher Columbus.

George Washington chose this site for the nation's capital. He also chose the engineer who designed the city, a Frenchman named Pierre L'Enfant. L'Enfant designed this grand city with many large avenues, parks, circles, and public buildings on high ground.

Washington is divided into four parts: the Northeast, Northwest, Southeast, and Southwest. The Capitol is the center of the city. Like many other towns, Washington has neighborhoods, schools, shops, and restaurants. There are brick and stone houses connected to each other in one long row. There are stately mansions and large, modern office buildings. Embassy Row features the homes of ambassadors from other countries.

Washington is home to many museums and galleries, several of which are part of the Smithsonian Institution. In addition to the Capitol and the national monuments, tourists can visit the Federal Bureau of Investigation (FBI) or the National Zoological Park. Tourists can even rent a bicycle and pedal through the Chesapeake and Ohio Canal, a national historic park that is located near the Potomac River.

From its beginnings as a rugged little town, Washington has become a magnificent city with many attractions. The best part of Washington, D.C., is that it belongs to everyone in the United States. It truly is America's hometown.

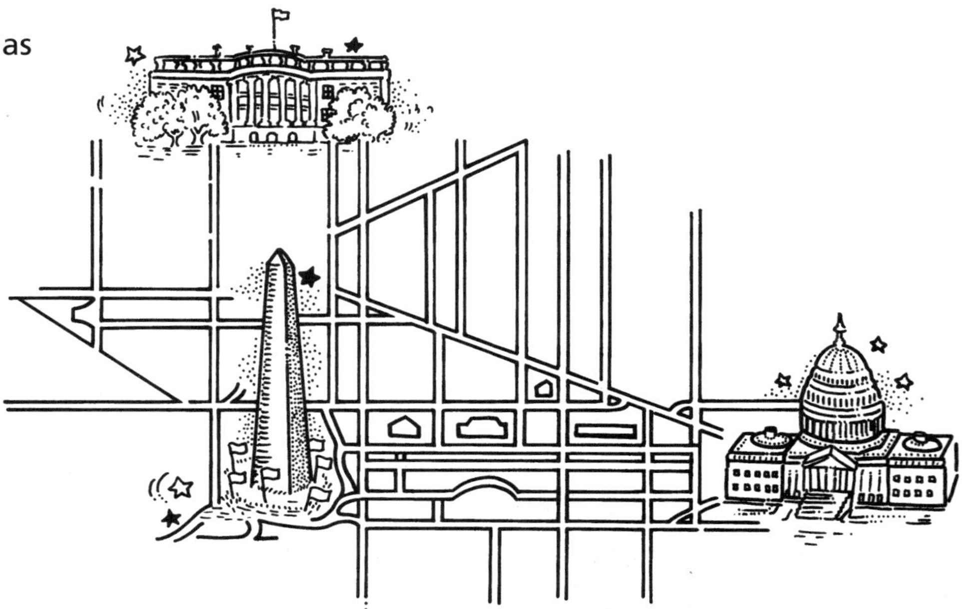

America's Hometown

Directions **Using what you have just read, answer the questions.**

1. How is Washington, D.C., an international city?

___

___

2. What does **D.C.** stand for?

___

3. What does the article mean when it says that Washington, D.C., is *a city without a state*?

___

___

4. How is Washington, D.C., like your hometown?

___

___

5. How is Washington, D.C., different from your hometown?

___

___

6. If you were to visit Washington, D.C., what would you like to see?

___

___

Sentence Types

REMEMBER

There are four kinds of sentences. A **declarative sentence** tells a fact or an opinion. It ends in a period. An **interrogative sentence** asks something. It ends with a question mark. An **imperative sentence** gives a direction, makes a request, or gives a command. It ends with either a period or an exclamation point. An **exclamatory sentence** shows strong feeling. It ends with an exclamation point.

EXAMPLES

Declarative: New York City is full of surprises.
Interrogative: Have you ever visited New York City?
Imperative: Look at the photograph of the Statue of Liberty.
Exclamatory: What an amazing city this is!

Name that Sentence

Directions Write the correct end punctuation marks for each sentence. Then write if the sentence is declarative, interrogative, imperative, or exclamatory.

1. The Statue of Liberty was a gift from France ____________________
2. Did you know that New York City has a subway system ____________________
3. These skyscrapers are huge ____________________
4. Come with us to the United Nations Building ____________________
5. The Metropolitan Museum is New York's largest museum ____________________
6. Did you see the costume exhibit at the museum ____________________
7. I could spend all day in Central Park ____________________
8. Look at those street performers ____________________
9. Millions of people visit New York City each year ____________________
10. I'm coming back as soon as I can ____________________

Historical Analogies

Directions **Complete the analogies using words from the box below.**

colonial	constitution	explore	frontier
homestead	immigrant	pioneers	historic

1. EMIGRANT is to LEAVE as ______________________ is to ARRIVE.
2. DECLARE is to DECLARATION as CONSTITUTE is to ______________________.
3. CREATE is to INVENT as TRAVEL is to ______________________.
4. WRITERS are to AUTHORS as ______________________ are to EXPLORERS.

Fathers of Our Nation

Directions **Read the report. Use words from the box above to fill in the blanks.**

Life was difficult during **(5)** ______________________ times. Many **(6)** ______________________, or early settlers, packed their belongings into wagons and headed West. The land of this new **(7)** ______________________ was often harsh and unforgiving. To build a **(8)** ______________________, these people had to clear the land and build the house using their hands. Today, many museums have relics and pictures that are **(9)** ______________________ reminders of those who began our nation. Their bravery to **(10)** ______________________ new lands is what built America.

Writing Fractions as Decimals

REMEMBER

To write a fraction as a decimal, divide the denominator into the numerator. This means you divide the bottom number into the top number.

EXAMPLE

Write $\frac{3}{4}$ as a decimal.

Write a division problem.

$4\overline{)3}$

Place a decimal point next to the 3. Write 0s to the right of the 3.

$4\overline{)3.00}$

Place a decimal point in the quotient. Divide.

$$\begin{array}{r} 0.75 \\ 4\overline{)3.00} \\ -2\ 8 \\ \hline 20 \\ -20 \\ \hline \end{array}$$

Directions Write each fraction as a decimal.

1. $\frac{7}{10} = 10\overline{)7.0} =$

2. $\frac{2}{5} = 5\overline{)2.0} =$

3. $\frac{1}{2} = 2\overline{)1.0} =$

4. $\frac{1}{8} =$

5. $\frac{3}{8} =$

6. $\frac{7}{40} =$

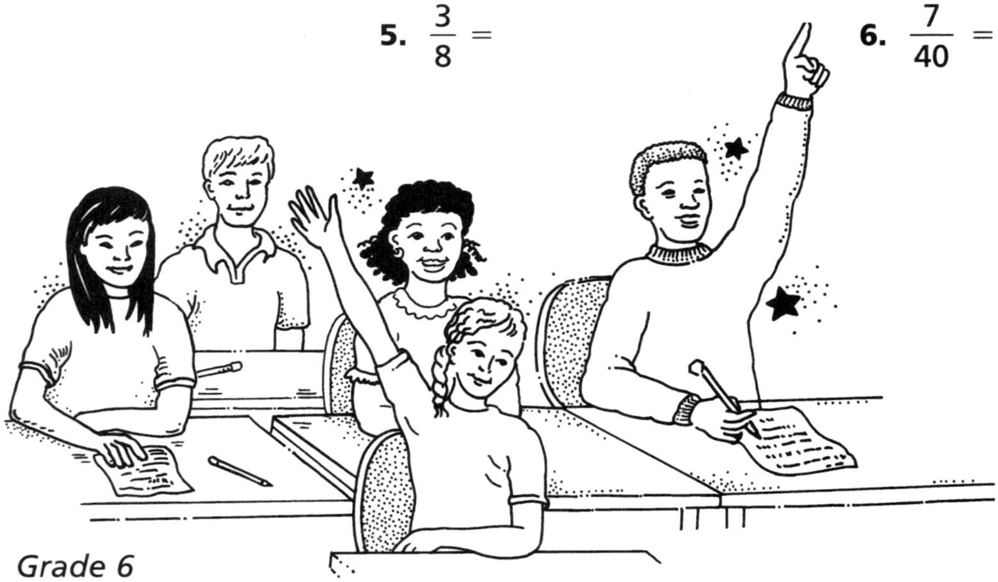

Find the Value

Directions Keisha divided the line segment $\overline{AB}$ in two ways. Use the drawing below to complete the page.

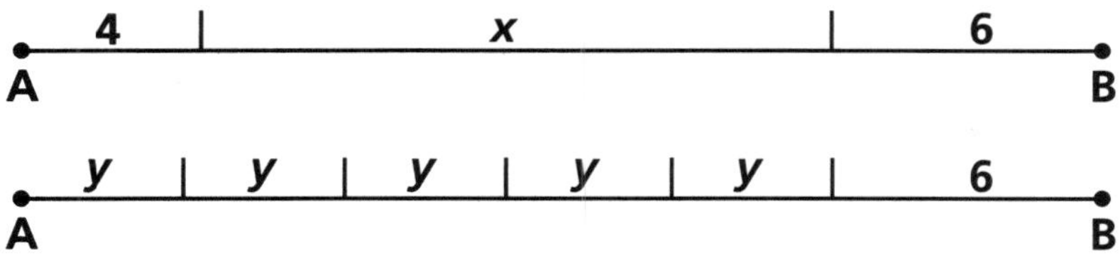

1. What is the value of x if $\overline{AB}$ is 24 feet long? ________ feet

2. What is the value of x if $\overline{AB}$ is 30 feet long? ________ feet

3. What is the value of y if $\overline{AB}$ is 26 feet long? ________ feet

4. What is the value of y if $\overline{AB}$ is 31 feet long? ________ feet

5. Circle the equation that shows the relationship between x and y.

   a. $x = \frac{1}{5}y - 4$

   b. $x = 5y + 4$

   c. $x = 5y - 4$

   d. $x = \frac{1}{5}y + 4$

Words About America

Directions Find the words listed in the box. Then circle them in the puzzle. The words can be hidden across, down, or diagonally.

country	patriot	Constitution	independence	president
proud	liberty	freedom	immigrant	democracy

F	P	D	E	M	O	C	R	A	C	Y	O
I	R	A	U	L	C	L	O	S	O	I	L
M	E	B	C	D	I	X	E	Y	N	W	U
M	S	P	X	Z	O	B	Q	V	S	S	C
I	I	A	A	S	A	P	E	W	T	R	O
G	D	D	F	T	P	S	R	R	I	Q	U
R	E	I	N	E	R	Y	G	O	T	W	N
A	N	N	Q	A	T	I	N	A	U	Y	T
N	T	F	R	E	E	D	O	M	T	D	R
T	N	E	T	Z	U	M	R	T	I	B	Y
R	C	O	R	U	L	M	R	I	O	N	D
I	N	D	E	P	E	N	D	E	N	C	E

What to Pack?

Directions Long ago, settlers traveling West had to carry everything with them in their Conestoga wagons, so they had to pack lightly. Imagine you and your family are traveling to a new frontier, but there isn't much room for your belongings. If you could only take one item with you, what would it be and why? Write about that item. Explain why you chose it.

CHAPTER 10

Choppers

Let's hear it for teeth! Squashers, crushers, choppers, grinders, chewers, and chompers—different shapes for different jobs. Without them, we'd have to drink our dinner!

Human teeth are designed to last a lifetime. They're covered with enamel, the hardest substance in the body. The enamel on your teeth can stand years of wear and tear, except for the chemical kind. Chemical wear and tear begins with bacteria.

Bacteria are tiny, one-celled organisms that live nearly everywhere. Some bacteria are helpful, but others are harmful and can cause disease. Bacteria love teeth. They feed on leftover pieces of food and make waste products, such as acids. The acids then attack your teeth. Dead bacteria, their waste, old food, and saliva can form a sticky layer of plaque, which combines with other substances to form tartar. Plaque and tartar build up and cause decay, gum disease, toothaches, and tooth loss.

With proper dental hygiene, you can prevent this buildup. Just follow the suggestions below for clean, healthy teeth and gums.

- Brush thoroughly at least twice a day. Floss once a day. You should brush for at least five minutes in order to completely clean all tooth surfaces.
- Use a brush with soft bristles. Hard bristles can harm the gums. As you brush, use enough pressure to feel the brush against your gums, but don't press so hard that you damage the gums.
- Use a fluoride (FLOR-ide) toothpaste. Fluoride is effective in helping teeth resist the acid that forms cavities.
- See your dentist twice a year for a cleaning and a tartar removal. (Your dentist may recommend that you visit more often.)
- Eat well-balanced meals that give you the nutrients (vitamins and minerals) your gums and teeth need.

Remember that unlike other parts of your body, your teeth cannot repair themselves if they are damaged or diseased. With good dental care, though, your teeth can last your entire lifetime.

Choppers

Directions Using what you have just read, answer the questions.

1. Why do people need teeth?

2. What does **decay** mean?

3. What does acid do to the teeth?

4. List two ways to keep teeth healthy.

5. How does eating good food help the teeth stay healthy?

6. What might happen to your teeth if you do not take care of them?

Compound Sentences

REMEMBER

A **simple sentence** has one complete subject and one complete predicate. This type of sentence may contain a compound subject, a compound predicate, or both. A **compound sentence** is formed when you join two or more simple sentences. Each simple sentence in a compound sentence is an **independent clause**. The independent clauses of a compound sentence are usually joined by the conjunction *and*, *but*, *or*, *nor*, and *yet*. Remember to use a comma before the conjunction in a compound sentence.

EXAMPLES

Simple sentence: *Ana* wants to eat a healthier diet.
Compound subject: *Ana* and *Brigid* shopped at a health food store.
Compound predicate: They *read food labels* and *asked many questions*.
Compound subject and compound predicate: *Ana* and *Brigid shopped* and *ate lunch*.

Simple or Compound?

Directions **Decide whether each sentence is simple or compound. Then write simple or compound on the lines.**

1. Good nutrition is important for everyone. ____________
2. It helps you grow and be healthy. ____________
3. Have your eating habits changed, or have they stayed the same? ____________
4. What foods do you eat for snacks? ____________
5. Ray doesn't like pizza that much, but he eats it because his friends do. ____________
6. Ray and his friends eat fast foods often. ____________
7. Nadia eats a mostly vegetarian diet. ____________
8. She doesn't eat beef, but she eats fish and chicken. ____________
9. Do you eat breakfast or do you skip it in the morning? ____________

Feeling Good Jumble

Directions **Unscramble the letters to make words from the box below.**

admired	grateful	confidence	enthusiasm	respect
generous	jubilant	enjoy	encouraged	

1. rtecsep ______________________

2. mesnatihsu ______________________

3. tailbunj ______________________

4. ddiarme ______________________

Student Chef

Directions **Read the letter. Use words from the box above to fill in the blanks.**

Dear Aunt Louise,

Thank you for the beautiful gift. That was very **(5)** ______________________ of you! I am **(6)** ______________________ for the gift and your **(7)** ______________________ in my cooking abilities. Part of the reason that I am going to learn to be a chef is because of the time we spent together. I have always **(8)** ______________________ the patience you had teaching me to cook. Your help **(9)** ______________________ me to always do my best. I'm sure I will **(10)** ______________________ cooking for others as much as you do. I just hope some day you will share all your secret recipes with me!

Love,

Lena

Percentages

REMEMBER

Percent means parts out of 100. The symbol **%** is used to show percent. Dividing a whole into 100 equal parts is an easy way to see percent.

EXAMPLE

25% means 25 parts out of 100.
25% is read **25 percent**.

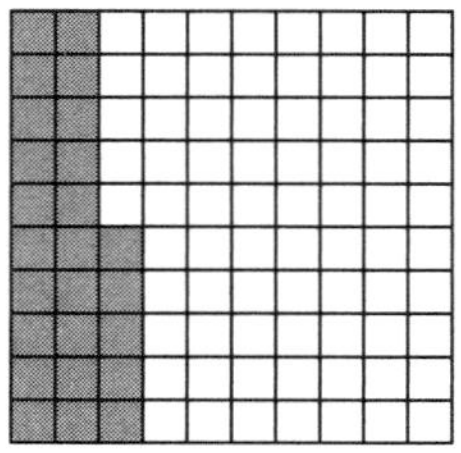

A quarter is 25% of a dollar.

A Healthy Percentage

Directions **In each problem, write the:**

- **percent of each grid that is shaded;**
- **percent of each grid that is not shaded; and**
- **sum (total) of the two percents.**

1. 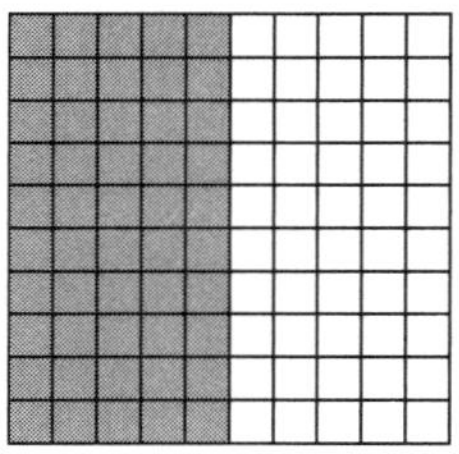

% shaded ______

% not shaded ______

total percent ______

2. 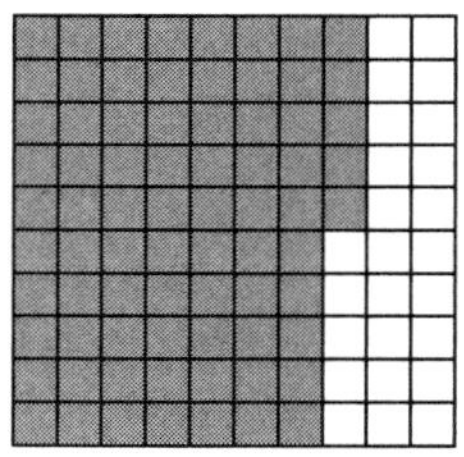

% shaded ______

% not shaded ______

total percent ______

3. 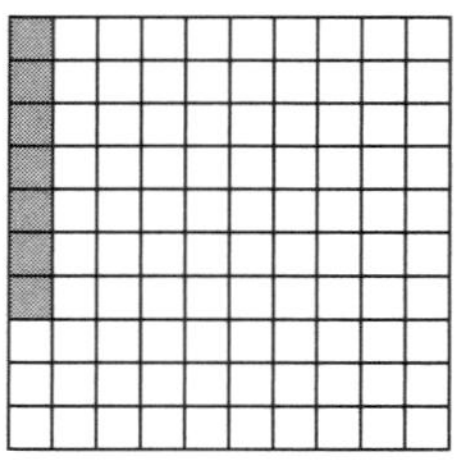

% shaded ______

% not shaded ______

total percent ______

4. 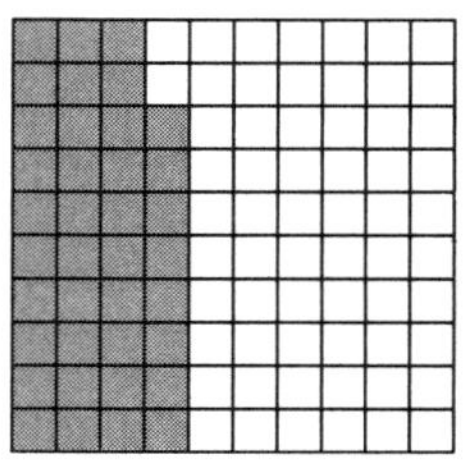

% shaded ______

% not shaded ______

total percent ______

What Are the Odds?

Directions Read the information below. Then complete the page.

Suppose a weather reporter says, "There's a 50% chance for rain tomorrow." The reporter is giving the probability for rain as a **percent**. A probability of 50% is the same as a probability of $\frac{1}{2}$.

Probability tells what is **most likely** to happen.

- If a probability is greater than 50% (or $\frac{1}{2}$), an event is **more likely to occur** than not to occur.
- If a probability is less than 50% (or $\frac{1}{2}$), an event is **more likely not to occur** than to occur.

1. There is a $\frac{1}{10}$ chance for sunshine tomorrow.

What is the chance in percent for sunshine tomorrow?

_______ %

2. There is a $\frac{3}{20}$ chance of rain tomorrow.

What is the chance in percent for rain tomorrow?

_______ %

3. There is a $\frac{3}{4}$ chance for snow tomorrow.

What is the chance in percent for snow tomorrow?

_______ %

4. If the information on this page is correct, what is most likely tomorrow: rain, sunshine, or snow? Explain your answer.

____

____

____

That's a Mouthful!

Directions Look at the codes. Each number stands for a letter. Write the words on the lines to answer the riddles.

A	B	C	D	E	F	G	H	I	J
1	2	3	4	5	6	7	8	9	10
K	**L**	**M**	**N**	**O**	**P**	**Q**	**R**	**S**	**T**
11	12	13	14	15	16	17	18	19	20
U	**V**	**W**	**X**	**Y**	**Z**				
21	22	23	24	25	26				

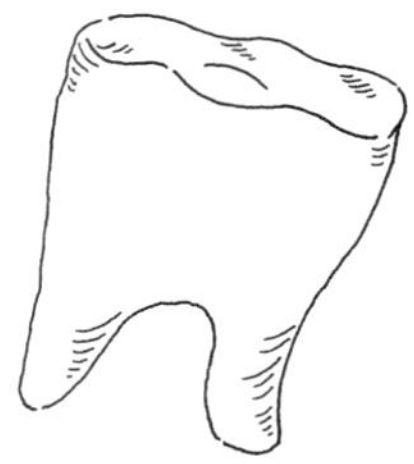

1. When are teeth like a train?

   23 8 5 14   20 8 5 25   3 8 5 23   3 8 5 23.

2. What did the gum say to the aching tooth?

   25 15 21 ' 22 5   7 15 20   19 15 13 5   14 5 18 22 5!

3. Why was the broken tooth so happy?

   14 15 23   8 5   23 1 19   1   3 8 9 16

   15 6 6   20 8 5   15 12 4   2 12 15 3 11.

4. Why did the tooth think it was a king?

   9 20   7 15 20   1   3 18 15 23 14.

5. Why did the girl want to go fishing?

   19 8 5   8 5 1 18 4   20 8 1 20   8 5 18

   4 1 4   8 1 4   1

   18 15 15 20   3 1 14 1 12.

Staying Healthy

Directions There are many things you can do to stay healthy, such as eating right and exercising. Create a daily schedule listing what you will do to stay healthy. Don't forget to include what foods you will eat and what activities you'll do!

READING CHECK-UP

Antarctica

Antarctica is considered to be the loneliest place on earth. It is the only continent with no native human population. The first explorers to set foot on Antarctica did not arrive until 1899. It was not until 1911 that the first explorer reached the South Pole. The only people who live in Antarctica are scientists and explorers from different countries. However, no one lives in Antarctica for more than eighteen months at a time.

Antarctica receives less than four inches of rain or snow each year. Still, it contains 70% of the world's fresh water. This fresh water is found frozen in its ice caps. The average temperature in Antarctica is $^{-}60°$ F. On July 21, 1983, the temperature in Antarctica dipped to $^{-}129°$F. This is the lowest temperature ever recorded in the world!

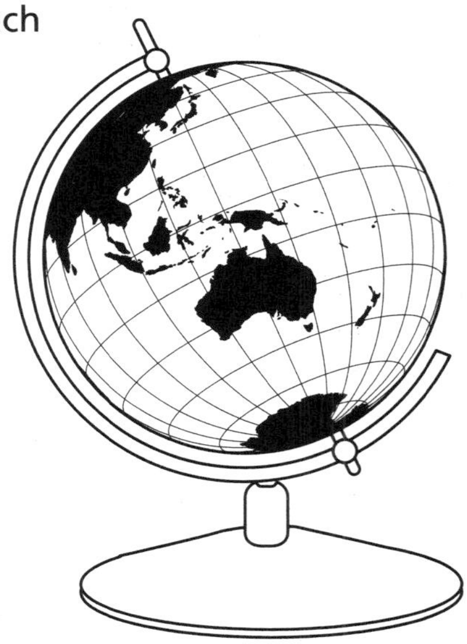

Besides being cold, Antarctica is also very windy. Winds blow from the South Pole toward the coasts for days and weeks without stopping. Because of the severe cold and wind, no animals live on the continent. The only native life includes a few plants, mosses, and some insects.

Directions **Fill in the bubble next to the correct answer.**

1. What is this article mostly about?

Ⓐ explorations to the South Pole
Ⓑ animals found in Antarctica
Ⓒ scientists who work in Antarctica
Ⓓ the continent of Antarctica

2. What happened in 1983?

Ⓐ The first explorers reached Antarctica.
Ⓑ The South Pole was discovered.
Ⓒ The lowest temperature in the world was recorded.
Ⓓ The first sign of insect life was found.

3. Animals can't live on Antarctica because:

Ⓐ there is no rain.
Ⓑ it is too cold and windy.
Ⓒ there are no plants to eat.
Ⓓ there is no water to drink.

4. The best summary for this article is:

Ⓐ Antarctica is the loneliest place on earth.
Ⓑ Antarctica has qualities unlike any other continent.
Ⓒ Antarctica is home to the South Pole.
Ⓓ Antarctica is filled with many species of plants and animals.

READING CHECK-UP

Not a Good Day

Jorge Castro's sister Rosie was shorter than most of the other girls in her class, but she was very fast and strong. In fact, she was the best athlete at her entire high school. Rosie was the star of the school's soccer and track teams. Jorge was not the best athlete and that was his problem. He knew his dad expected him to be a sports star, too.

It was a lovely spring afternoon of Jorge's first year in middle school. It was the time of year that he loved most. He liked that he could get outside on his bike and explore the fields and woods around the small town where the Castros lived. On this particular afternoon, though, Jorge was not his usual happy self as he walked the five blocks from school to his house. He had just learned that he didn't make the school baseball team. Now he had to go home and tell his father that all their hours of practice in the backyard hadn't been enough. At home, Jorge went immediately to his room and closed the door, hoping to delay delivering the bad news for as long as possible.

Finally, as Jorge slid into his place at the supper table that night, the dreaded question came. "How did the try-outs go today, Jorge?" asked Mr. Castro. "Did the coach put you on first base?"

"No," Jorge said. He looked down at his plate and said in a softer voice, "I didn't make the team."

"You'll make it next time," said Mr. Castro. "At least now you have more time to go exploring on your bike." Jorge looked up at his dad. They both smiled.

Directions Fill in the bubble next to the correct answer.

5. Why was Rosie the best athlete at her high school?

- (A) She was older than her brother.
- (B) Rosie was shorter than most of the other girls in her class.
- (C) Rosie was very fast and strong.
- (D) Her father had coached her for years.

6. Jorge Castro liked spring because

- (A) he could play baseball.
- (B) he was now in middle school.
- (C) he could spend time with his father.
- (D) he could explore the woods and fields on his bike.

7. After Jorge learned he didn't make the baseball team, he

- (A) went to watch his sister practice.
- (B) went to his room.
- (C) rode his bike into the woods.
- (D) practiced in the backyard.

8. Why did Jorge go to his room?

- (A) He wanted to do his homework.
- (B) He wanted to avoid his father.
- (C) He wanted to make a phone call.
- (D) He wanted to take a nap.

READING CHECK-UP

The Industrial Revolution

Most people think of a revolution as a war between kings and common people. During the American and French Revolutions, common people revolted against their kings and queens. Then they selected the person they wanted as their leader. You can imagine these revolutions brought about big changes. Other revolutions brought changes in ideas and technology. In England during the late 1750s, the Industrial Revolution began. The advances in technology made during this period affected many lives.

If a European peasant living in the 1500s were magically transported to the year 1750, he would not have found life very different. Most people still lived in the country and worked on the land. People made most of their own goods. Their homes, clothing, and furniture were all made by hand. Few people traveled very far. Most never left the counties where they were born. When they did travel, they went by foot or horse. If the same peasant arrived in the 1800s, he would have found himself in a very different world.

During the Industrial Revolution, thousands of people left the country to find jobs in the cities. The population of cities grew so quickly that there weren't enough houses for everyone. In factories, a single machine could take the place of hundreds or thousands of people. The houses people lived in, the furniture they used, and the clothes they wore were all made by others. Most people now bought food that was grown and harvested by someone else. More and more people traveled. Thousands moved across the country or across oceans to find work. Railroads and steamships replaced horses and wagons. In under a decade or two, many lives in England almost completely changed.

Directions **Fill in the bubble next to the correct answer.**

9. The main idea of this article is:

- (A) The Industrial Revolution brought an end to the Middle Ages.
- (B) The American and French Revolutions brought changes.
- (C) The Industrial Revolution changed life in Europe.
- (D) Many people traveled during the Industrial Revolution.

11. In the early 1750s, people in Europe

- (A) traveled by railroad.
- (B) lived in rural areas.
- (C) moved to the cities to find work.
- (D) worked in factories.

10. The Industrial Revolution began

- (A) before 1500.
- (B) between 1500 and 1750.
- (C) between 1750 and 1800.
- (D) after 1800.

12. During the Industrial Revolution, people

- (A) used more manufactured goods.
- (B) made the goods they needed by hand.
- (C) stopped traveling far from home for work.
- (D) had larger families.

READING CHECK-UP

Opening Night

"Ten minutes until the opening act," announced Mr. Ruiz. "Is everyone ready for the performance?"

Beads of sweat began to form on Tanya's forehead. The knot in her stomach began to tighten. Tanya had never felt this nervous before.

"I can do this," Tanya repeated to herself. "We've rehearsed the play for six weeks, and I memorized all my lines and cues. This anxiety will go away."

Tanya was having a sudden attack of stagefright, but she couldn't understand why. She had been in every class play since first grade, making this her seventh performance.

"What's the matter, Tanya?" asked her best friend, Amy Wong. "You look nauseous."

"My stomach is upset and my knees won't stop shaking!" replied Tanya.

"It's just nerves," said Amy. Amy knew all about stagefright. She was in the same grade as Tanya, but belonged to the Community Theater Group. She had acted in more than twenty plays in the last three years. "Take a few deep breaths and count to three. Once the curtain opens, everything will be just like we practiced."

"I hope so," Tanya said, her face growing pale. She bit her lip and took her position on stage with the other actors.

"We're on," Mr. Ruiz whispered from offstage as the curtain began to open.

Tanya stood for a moment and looked out at the crowd gathered in the school auditorium. The music began to play, the spotlights beamed down on her, and the first actor began to speak his lines. Suddenly, Tanya began to feel different. Her stomach relaxed and her knees were no longer shaking. She stood up straight and took a deep breath. She was prepared to deliver her lines.

Directions **Fill in the bubble next to the correct answer.**

13. The play is being performed

Ⓐ in a classroom.

Ⓑ in a community theater.

Ⓒ in a gymnasium.

Ⓓ in the school auditorium.

14. What will probably happen next?

Ⓐ Tanya will run off stage.

Ⓑ Tanya will quit acting.

Ⓒ Tanya will deliver her lines.

Ⓓ Tanya will forget her lines.

15. Which statement expresses a fact?

Ⓐ Amy belongs to a theater group.

Ⓑ Mr. Ruiz should have held more play rehearsals.

Ⓒ Tanya had no reason to feel nervous.

Ⓓ Amy has more talent than Tanya.

16. The author wrote this story to

Ⓐ entertain readers with a story about a school play.

Ⓑ inform readers about how to deal with being nervous.

Ⓒ describe how most actors feel on opening night.

Ⓓ convince readers to act in a school play.

Stop! **Number Correct:** ______ **out of 16**

MATH CHECK-UP

Directions **Read each question. Fill in the bubble next to the correct answer.**

1. The 6 in 345,697 means

Ⓐ 6,000
Ⓑ 60,000
Ⓒ 600,000
Ⓓ 600

2. $500{,}000 + 30{,}000 + 200 + 80 =$

Ⓐ 500,030,280
Ⓑ 532,080
Ⓒ 530,280
Ⓓ 5,328

3. Which number would go in the box to make this number sentence true?

$3 \times (2 + 7) = (2 + 7) \times \square$

Ⓐ 3
Ⓑ 6
Ⓒ 21
Ⓓ 42

4. What two numbers are missing from the pattern?

3, 7, 11, _____, _____, 23, 27

Ⓐ 12, 22
Ⓑ 14, 18
Ⓒ 15, 19
Ⓓ 13, 17

5. Which number is a common factor of 12 and 27?

Ⓐ 2
Ⓑ 3
Ⓒ 4
Ⓓ 9

6. Which of these expresses 75% as a fraction?

Ⓐ $\frac{75}{1000}$
Ⓑ $\frac{75}{100}$
Ⓒ $\frac{75}{10}$
Ⓓ $\frac{75}{1}$

7. Which represents $\frac{81}{1000}$ as a decimal?

Ⓐ 0.0081
Ⓑ 8.01
Ⓒ 0.81
Ⓓ 0.081

8. Which number is represented by the *x* on the number line?

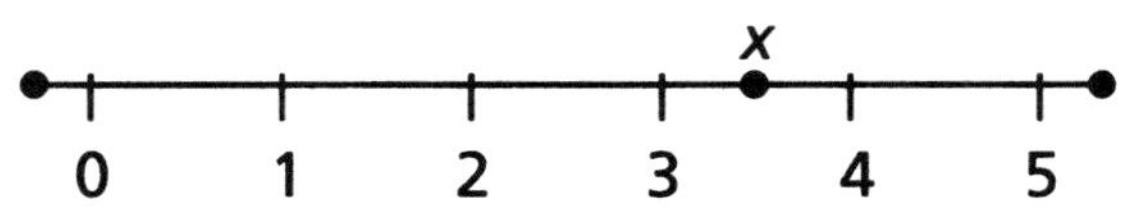

Ⓐ 3.2
Ⓑ 3.5
Ⓒ 3.8
Ⓓ 4.5

MATH CHECK-UP

Directions Read each question. Fill in the bubble next to the correct answer.

9. $\begin{array}{r} 70{,}282 \\ -\ 1{,}395 \\ \hline \end{array}$

Ⓐ 6887 Ⓑ 68,887
Ⓒ 68,897 Ⓓ 69,887

10. $\begin{array}{r} 509 \\ \times\ 12 \\ \hline \end{array}$

Ⓐ 521 Ⓑ 6,118
Ⓒ 6,018 Ⓓ 6,108

11. $7\overline{)5341}$

Ⓐ 863 Ⓑ 773
Ⓒ 763 Ⓓ 767 R3

12. $0.625 + 0.437 =$

Ⓐ 1.062 Ⓑ 0.1062
Ⓒ 1.061 Ⓓ 1.072

13. $0.6 \times 0.7 =$

Ⓐ 4.2 Ⓑ 0.42
Ⓒ 0.042 Ⓓ 0.36

14. $3.67 \div 10 =$

Ⓐ 0.0376 Ⓑ 0.367
Ⓒ 36.7 Ⓓ 3.67

15. $0.352 - 0.173 =$

Ⓐ 0.179 Ⓑ 0.189
Ⓒ 0.757 Ⓓ 1.79

16. $\frac{2}{15} + \frac{4}{45} =$

Ⓐ $\frac{6}{60}$ Ⓑ $\frac{8}{45}$
Ⓒ $\frac{6}{75}$ Ⓓ $\frac{10}{45}$

17. $\begin{array}{r} \frac{1}{8} \\ +\ \frac{5}{16} \\ \hline \end{array}$

Ⓐ $\frac{5}{128}$ Ⓑ $\frac{6}{24}$
Ⓒ $\frac{7}{16}$ Ⓓ $\frac{6}{16}$

18. $\begin{array}{r} \frac{2}{3} \\ -\ \frac{2}{6} \\ \hline \end{array}$

Ⓐ $\frac{1}{3}$ Ⓑ $\frac{3}{6}$
Ⓒ $\frac{4}{9}$ Ⓓ $\frac{4}{18}$

MATH CHECK-UP

Directions **Read each question. Fill in the bubble next to the correct answer.**

The graph shows popular pizza toppings among middle school students. Use the graph to answer questions 19, 20, and 21.

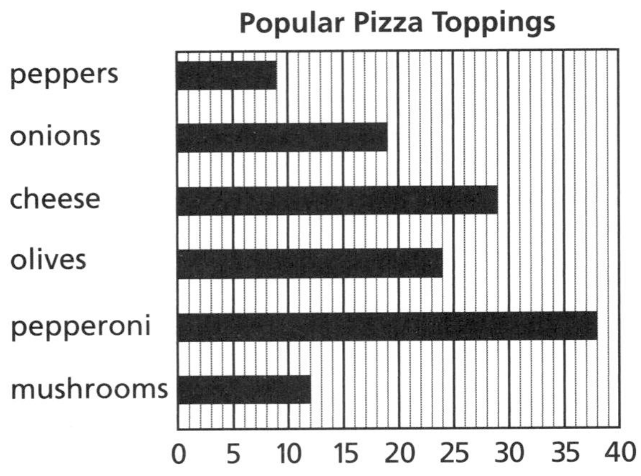

19. Which topping is more popular than cheese?

Ⓐ onions Ⓑ olives
Ⓒ mushrooms Ⓓ pepperoni

20. Which two toppings are the least popular?

Ⓐ peppers and olives
Ⓑ mushrooms and pepperoni
Ⓒ peppers and mushrooms
Ⓓ cheese and onions

21. About how many more students prefer cheese than peppers as a topping?

Ⓐ 20 Ⓑ 25
Ⓒ 30 Ⓓ 35

22. Rita has 12 shirts. Half of them have long sleeves. Half of the long-sleeved shirts are striped. How many shirts are striped?

Ⓐ 2 Ⓑ 3
Ⓒ 4 Ⓓ 9

23. A theater is offering a special price of $24.75 for 7 movie tickets. About how much does each ticket cost at this special price?

Ⓐ $31.75 Ⓑ $3.50
Ⓒ $3.00 Ⓓ $3.75

24. Juan puts 8 marbles in a bag. Three marbles are red, 2 marbles are green, 1 marble is yellow, and the remaining marbles are blue. What are the chances Juan will reach in the bag and pull out a blue marble?

Ⓐ $\frac{1}{4}$ Ⓑ $\frac{3}{8}$
Ⓒ $\frac{1}{8}$ Ⓓ $\frac{1}{2}$

25. For 4 weeks, the Cruz's food bills were $125, $87, $132, and $92. What was the average cost of food per week?

Ⓐ $109 Ⓑ $436
Ⓒ $190 Ⓓ $100

Stop! **Number Correct:** ______ **out of 25**

Congratulations!

(name)

has completed *Summer Counts!*

Good job!

Have a good school year.

ANSWER KEY

Page 5

1. It is about Keisha and her first job.
2. Keisha is afraid she won't know what to do.
3. Her mother asks her to work at her aunt's restaurant.
4. Possible answer: Keisha did a good job on her first night.
5. Possible answer: The author wants to tell what it's like to start a new job.
6. Possible answer: Keisha will take dirty dishes to Eddie.

Page 6

1. her
2. She
3. his
4. their
5. them
6. their
7. I
8. them
9. my
10. they

Page 7

1. guarantee
2. savings
3. value
4. inexpensive
5. inherit
6. Financial
7. income
8. wage
9. savings
10. withdraw

Page 8

1. $\frac{1 \times 1}{2} = \frac{1}{2}$
2. $\frac{2 \times 1}{4} = \frac{2}{4} = \frac{1}{2}$
3. $\frac{2 \times 2}{4} = \frac{4}{4} = 1$
4. $\frac{2 \times 3}{3} = \frac{6}{3} = 2$
5. $\frac{3 \times 8}{4} = \frac{24}{4} = 6$
6. $\frac{5 \times 5}{12} = \frac{25}{12} = 2\frac{1}{12}$

Page 9

1. **a.** 41
   b. 57
   c. 14
   d. 37
   e. 13
2. **a.** \$20.50
   b. \$51.30
   c. \$17.50
   d. \$31.45
   e. \$19.50
3. Small Juice
4. \$1.00
5. Possible answer: No, a larger drink won't sell well. Higher priced drinks didn't sell well.

Page 10

1. withdraw
2. job
3. bank
4. dollar
5. money
6. save
7. spend
8. coins
9. tax

Riddle: He only had one scent!

Page 13

1. E.B. White wrote *Charlotte's Web*.
2. White wrote for *The New Yorker* magazine.
3. Possible answer: He watched the animals on his farm.
4. He felt spiders had gotten a bad reputation they didn't deserve.
5. Answers will vary. Accept reasonable responses.
6. Possible answer: White loved animals so he wrote about them.

Page 14

1. Beverly Cleary, books, children, adults
2. awards
3. Cleary, problems, pleasures
4. People, Cleary, ideas, books
5. Cleary, memories, childhood
6. tooth
7. friendships
8. letters, readers
9. Cleary, people, books
10. success

Page 15

1. journal
2. deadline
3. rhymes
4. edit
5. essay
6. topic
7. persuade
8. opinion
9. read
10. conclusion

Page 16

1. $\frac{2}{7}$
2. $\frac{1}{7}$
3. $\frac{3}{7}$
4. 0

Page 17

1. $\frac{2}{4}$ or $\frac{1}{2}$
2. $\frac{2}{4}$ or $\frac{1}{2}$
3. c; There are more blue marbles than red in the bag.

Page 18

Across

2. poem
6. biography
7. letter
8. haiku
9. myth
10. folktale

Down

1. journal
3. report
4. limerick
5. mystery

Page 21

1. The seven suns ruled the skies.
2. Possible answer: They asked Erh-Lang for help because he was a very strong giant.
3. Possible answers: The old woman went to Erh-Lang for help; she said to hide the suns in the mountain; she asked the birds to sing to the seventh sun.
4. The seventh sun hid and wouldn't come out.
5. Possible answer: Spite means to make someone angry.
6. Possible answer: The people and plants would die without food or water.

Page 22

1. Who
2. What
3. Who
4. Which
5. Who
6. What
7. Whose; Which
8. whom
9. Which
10. What

Page 23

1. planet
2. Mars
3. telescope
4. Astronomy
5. satellites
6. revolve
7. gravity
8. galaxy
9. constellation
10. planet

Page 24

1. $\frac{1 \times 1}{2 \times 2} = \frac{1}{4}$

2. $\frac{1 \times 4}{2 \times 5} = \frac{4}{10} = \frac{2}{5}$
3. $\frac{1 \times 3}{3 \times 5} = \frac{3}{15} = \frac{1}{5}$
4. $\frac{2 \times 1}{3 \times 2} = \frac{2}{6} = \frac{1}{3}$
5. $\frac{3 \times 2}{4 \times 5} = \frac{6}{20} = \frac{3}{10}$
6. $\frac{5 \times 4}{8 \times 6} = \frac{20}{48} = \frac{5}{12}$

Page 25

1. No, Mani only has $1\frac{1}{12}$ lbs. of potatoes.
2. a. 14 servings $(4 \times 3\frac{1}{2})$
   b. $4\frac{13}{16}$ lbs. potatoes; $\frac{7}{8}$ cups cheese; $1\frac{1}{6}$ cups broth; and $2\frac{5}{8}$ teaspoons salt
3. a. $\frac{11}{32}$ lb.
   b. 5

Page 26

M	I	M	U	I	P	R	L	V	M
C	E	B	A	E	Y	L	Q	S	U
D	S	R	T	R	Y	P	U	A	S
J	N	E	C	P	S	E	N	T	I
U	R	A	N	U	S	Y	E	U	O
P	T	R	I	A	R	S	P	R	V
I	U	T	H	X	W	Y	T	N	K
T	N	H	S	R	U	M	U	P	T
E	R	I	A	U	V	E	N	U	S
R	L	R	H	Y	N	H	E	E	T

Page 29

1. It had two wheels, a cross bar, and a carved horse's head to hold on to.
2. Possible answer: Bicycles today have pedals, brakes, gears, and handlebars.
3. Possible answer: Cycling is good exercise.
4. Possible answer: People know about bicycles, so they don't seem strange.
5. Possible answer: A mountain bike or BMX bike would be best.
6. Bicycles are used for travel, recreation, and sports.

Page 30

Possible answers:

1.–6. Answers will vary. Accept reasonable responses.

Page 31

1. bicycle
2. sleigh
3. ferry
4. airport
5. motorcycle
6. Taxicab
7. ferry
8. subway
9. airport
10. spacecraft

Page 32

1. $\frac{1}{4}$
2. $\frac{1}{3}$
3. $\frac{1}{2}$
4. $\frac{1}{5}$
5. $\frac{1}{4}$
6. $\frac{1 \times 1}{4 \times 3} = \frac{1}{12}$
7. $\frac{2 \times 1}{3 \times 4} = \frac{2}{12}$ or $\frac{1}{6}$
8. $\frac{3 \times 1}{8 \times 3} = \frac{3}{24}$ or $\frac{1}{8}$
9. $\frac{8 \times 1}{12 \times 2} = \frac{8}{24}$ or $\frac{1}{3}$
10. $\frac{9 \times 1}{12 \times 3} = \frac{9}{36}$ or $\frac{1}{4}$

Page 33

1. $\frac{3}{8}$
2. 6
3. $\frac{5}{8}$
4. 10
5. a. $\frac{1}{4}$
   b. 4
   c. Answers will vary. Accept reasonable responses.

Page 34

1. stuck truck
2. tractor actor
3. plain plane
4. train strain
5. ship trip
6. new canoe

Page 37

1. It made his art more realistic.
2. Possible answer: Verrocchio didn't like it because he was the teacher.
3. Possible answer: Yes. He became a better painter.
4. He didn't have the tools or materials needed to make them.
5. Answers will vary. Accept reasonable responses.
6. Possible answer: The author wanted to tell about Leonardo da Vinci and his art.

Page 38

1. adjective
2. adjective
3. adverb
4. adjective
5. adjective
6. adverb
7. adjective
8. adverb

Page 39

1. collage
2. charcoal
3. original
4. canvas
5. scarlet
6. sketch
7. exhibit
8. Gallery
9. original
10. portrait
11. canvas
12. sculpture

Page 40

1. $\frac{4}{8} = \frac{1}{2}$
2. $\frac{6}{10} = \frac{3}{5}$
3. $\frac{15}{20} = \frac{3}{4}$
4. $\frac{8}{6} = \frac{4}{3}$
5. $\frac{10}{6} = \frac{5}{3}$
6. $\frac{20}{5} = \frac{4}{1}$
7. $\frac{12}{4} = \frac{3}{1}$
8. $\frac{8}{4} = \frac{2}{1}$
9. $\frac{50}{10} = \frac{5}{1}$
10. $\frac{80}{4} = \frac{20}{1}$
11. $\frac{18}{16} = \frac{9}{8}$
12. $\frac{81}{9} = \frac{9}{1}$

Page 41

1. a. $6\frac{3}{4} \div n = \frac{3}{4}$
   b. 9
2. a. $p \div 1\frac{1}{2} = 6$
   b. 9
3. a. Possible answer: $c = 6\frac{7}{8} \div 11$
   b. $\frac{5}{8}$ lb.

Page 42

1. She wanted to brush up on her strokes.
2. The clues were sketchy.
3. I am picture perfect!
4. He was quick on the draw.

Page 45

1. Possible answer: It is about the Roman aqueducts.
2. Aqueducts are plumbing systems that carry water from the mountains into Rome.
3. Possible answer: Gravity helped move the water downhill.
4. Possible answer: Unravel means to discover.
5. Answers will vary. Accept reasonable responses.
6. Possible answers: The flight was pleasant, but the car ride was more interesting; I am fascinated by this ancient city.

Page 46

1. at the camp, counselors

2. of our counselor, name
3. about Camp Arrowhead, stories
4. of Penny, picture
5. near the cabin, girls
6. of them, Two
7. from many states, Girls
8. at the camp, food
9. near the camp, Gardens
10. below the camp, river

Page 47

1. takeoff
2. tourist
3. transportation
4. excursion
5. vacation
6. exciting
7. luggage
8. gasoline
9. gear
10. motel
11. tourist

Page 48

1. 0.005
2. 0.06
3. 0.04
4. 0.078
5. 0.109
6. 0.253
7. six thousandths
8. seven tenths
9. thirteen hundredths
10. three and seventy-seven thousandths

Page 49

1. 1st: Mickey
   2nd: Gerald
   3rd: Gunnar
   4th: Darrell
   5th: Nick
2. 1.475 seconds
3. Nick and Darrell
4. 0.513 seconds
5. 69.073 seconds
6. 69.221 seconds

Page 50

1. subway
2. postcard
3. luggage
4. airport
5. takeoff
6. pack
7. travel

Riddle: A lawsuit case

Page 53

1. He could make himself invisible.
2. She helped him do great deeds; She helped him find a wife.
3. They were jealous of her.
4. She patched her clothes and put on the few ornaments she had.
5. Glenna answered the question honestly.
6. Possible answer: A girl's older sisters treat her badly, but she wins in the end.

Page 54

1. Wow!
2. Gee
3. Hey
4. Yikes!
5. Oops
6. Shh
7. Oh, no!
8. Goodness
9. Oh, my!
10. Great!

Page 55

1. blizzard
2. lightning
3. tornado
4. drought
5. blizzard
6. thunderstorms
7. lightning
8. hurricane
9. sandstorms
10. drought
11. weather

Page 56

1. 1.2
2. 0.6
3. 0.8
4. 11.6
5. 18.5
6. 16.30
7. $0.04
8. $3.60
9. $5.32

Page 57

1. a. 0.01; $\frac{1}{100}$
   b. 0.05; $\frac{5}{100}$ or $\frac{1}{20}$
   c. 0.10; $\frac{10}{100}$ or $\frac{1}{10}$
   d. 0.25; $\frac{25}{100}$ or $\frac{1}{4}$
2. a. 0.23; $\frac{23}{100}$
   b. 0.29; $\frac{29}{100}$
   c. 1.50; $1\frac{50}{100}$ or $1\frac{1}{2}$
   d. 3.21; $3\frac{21}{100}$

Page 58

1. hail pail
2. thunder wonder
3. plain rain
4. drizzle sizzle
5. cloud crowd

Page 61

1. It is about studying earthquakes.
2. Possible answer: He can be alerted when a quake occurs.
3. Possible answers: The scientist wouldn't know a quake happened.
4. Possible answer: Reporters can warn others about them.
5. A machine that records the motion of the ground during an earthquake.
6. Possible answer: They could make buildings stronger and safer.

Page 62

1. Many scientists / are studying

Subject: scientists
Predicate: are studying

2. The fire ant / is

Subject: fire ant
Predicate: is

3. This ant / has

Subject: ant
Predicate: has

4. Fire ants / have invaded

Subject: ants
Predicate: have invaded

5. They / are slowly creeping

Subject: They
Predicate: are creeping

6. Their stings / can be

Subject: stings
Predicate: can be

7. The bug / feasts and destroys

Subject: bug
Predicate: feasts, destroys

8. The ant / has

Subject: ant
Predicate: has

Page 63

1. ecology
2. research
3. pollute
4. evaporate
5. research
6. technology
7. environment
8. biology
9. chemistry
10. ecology

Page 64

1. 0.2
2. 0.02
3. 0.03
4. 0.32
5. 0.31
6. 2.38
7. $0.84
8. $1.65
9. $7.92

Page 65

1. $7.2 + 0.8 = 8$
2. $1.6 - 0.6 = 1$
3. $3.6 + 3.1 = 6.7$
4. $7.2 + 4.8 = 12$
5. $5.68 - 4.68 = 1$

Page 66

Across

1. seismology

4. chemistry
6. geology
7. archeology
8. biology
9. botany
10. physics

Down

2. meteorology
3. ecology
5. zoology

Page 69

1. Possible answer: People from different nations have helped build it and live there.
2. It stands for District of Columbia.
3. Possible answers: Washington, D.C., isn't located in a state. It's inside a federal district.
4. Possible answer: It has neighborhoods, schools, shops, and restaurants.
5. Answers will vary. Accept reasonable responses.
6. Answers will vary. Accept reasonable responses.

Page 70

1. **.**; declarative
2. **?**; interrogative
3. **!**; exclamatory
4. **.**; imperative
5. **.**; declarative
6. **?**; interrogative
7. **!**; exclamatory
8. **.**; imperative
9. **.**; declarative
10. **!**; exclamatory

Page 71

1. immigrant
2. constitution
3. explore
4. pioneers
5. colonial
6. pioneers
7. frontier
8. homestead
9. historic
10. explore

Page 72

1. 0.7
2. 0.4
3. 0.5
4. $8\overline{)1.000} = 0.125$
5. $8\overline{)3.000} = 0.375$
6. $40\overline{)7.000} = 0.175$

Page 73

1. 14
2. 20
3. 4
4. 5
5. c

Page 74

Page 77

1. Possible answer: People need teeth to chew their food.
2. Decay means to rot.
3. Acid attacks teeth.
4. Possible answers: Brush at least twice a day; Floss daily; Use a brush with soft bristles; Use a fluoride toothpaste; See your dentist twice a year; Eat well-balanced meals.
5. Possible answer: Food has nutrients teeth need to stay strong.
6. Possible answers: You might get decay, gum disease, toothaches, or tooth loss.

Page 78

1. simple
2. compound
3. compound
4. simple
5. compound
6. compound
7. simple
8. compound
9. compound

Page 79

1. respect
2. enthusiasm
3. jubilant
4. admired
5. generous
6. grateful
7. confidence
8. admired
9. encouraged
10. enjoy

Page 80

1. 50%; 50%, 100%
2. 75%; 25%; 100%
3. 7%; 93%; 100%
4. 38%; 62%; 100%

Page 81

1. 10%
2. 15%
3. 75%
4. Snow is more likely. There's a higher probability for snow.

Page 82

1. When they chew chew.
2. You've got some nerve!
3. Now he was a chip off the old block.
4. It got a crown.
5. She heard that her dad had a root canal.

Page 84

1. D
2. C
3. B
4. B

Page 85

5. C
6. D
7. B
8. B

Page 86

9. C
10. C
11. B
12. A

Page 87

13. D
14. C
15. A
16. C

Page 88

1. D
2. C
3. A
4. C
5. B
6. B
7. D
8. B

Page 89

9. B
10. D
11. C
12. A
13. B
14. B
15. A
16. D
17. C
18. A

Page 90

19. D
20. C
21. A
22. B
23. B
24. A
25. A